NATIONAL ACADEMIES

Sciences
Engineering
Medicine

NATIONAL ACADEMIES PRESS
Washington, DC

Compounding Disasters in Gulf Coast Communities 2020–2021

Impacts, Findings, and Lessons Learned

Committee on
Compounding Disasters in
Gulf Coast Communities,
2020–2021:
Impacts, Findings, and
Lessons Learned

Gulf Health and Resilience Board

Gulf Research Program

Consensus Study Report

NATIONAL ACADEMIES PRESS 500 Fifth Street, NW Washington, DC 20001

This activity was supported by the Gulf Research Program of the National Academy of Sciences. Any opinions, findings, conclusions, or recommendations expressed in this publication do not necessarily reflect the views of any organization or agency that provided support for the project.

International Standard Book Number-13: 978-0-309-70716-9
International Standard Book Number-10: 0-309-70716-1
Digital Object Identifier: https://doi.org/10.17226/27170
Library of Congress Control Number: 2024945140

This publication is available from the National Academies Press, 500 Fifth Street, NW, Keck 360, Washington, DC 20001; (800) 624-6242; http://www.nap.edu.

Printed in the United States of America.

Suggested citation: National Academies of Sciences, Engineering, and Medicine. 2024. *Compounding Disasters in Gulf Coast Communities, 2020–2021: Impacts, Findings, and Lessons Learned*. Washington, DC: National Academies Press. https://doi.org/10.17226/27170.

The **National Academy of Sciences** was established in 1863 by an Act of Congress, signed by President Lincoln, as a private, nongovernmental institution to advise the nation on issues related to science and technology. Members are elected by their peers for outstanding contributions to research. Dr. Marcia McNutt is president.

The **National Academy of Engineering** was established in 1964 under the charter of the National Academy of Sciences to bring the practices of engineering to advising the nation. Members are elected by their peers for extraordinary contributions to engineering. Dr. John L. Anderson is president.

The **National Academy of Medicine** (formerly the Institute of Medicine) was established in 1970 under the charter of the National Academy of Sciences to advise the nation on medical and health issues. Members are elected by their peers for distinguished contributions to medicine and health. Dr. Victor J. Dzau is president.

The three Academies work together as the **National Academies of Sciences, Engineering, and Medicine** to provide independent, objective analysis and advice to the nation and conduct other activities to solve complex problems and inform public policy decisions. The National Academies also encourage education and research, recognize outstanding contributions to knowledge, and increase public understanding in matters of science, engineering, and medicine.

Learn more about the National Academies of Sciences, Engineering, and Medicine at **www.nationalacademies.org**.

COMMITTEE ON COMPOUNDING DISASTERS IN GULF COAST COMMUNITIES, 2020–2021: IMPACTS, FINDINGS, AND LESSONS LEARNED

ROY E. WRIGHT (*Chair*), President and Chief Executive Officer, Insurance Institute for Business and Home Safety
JEFF BYARD, Vice President of Operations, Team Rubicon
CRAIG COLTEN, Professor of Geography, Louisiana State University (Emeritus)
TRACY KIJEWSKI-CORREA, William J. Pulte Director, Pulte Institute for Global Development, Keough School of Global Affairs, University of Notre Dame; Professor of Engineering and Global Affairs; Academic Director, Integration Lab
J. MARSHALL SHEPHERD, Georgia Athletic Association Distinguished Professor; Director, Atmospheric Sciences Program, University of Georgia
JAMES M. SHULTZ, Associate Professor, Epidemiology Population Health Sciences, University of Miami Miller School of Medicine
CHAUNCIA WILLIS-JOHNSON, Chief Executive Officer, Institute for Diversity and Inclusion in Emergency Management

Study Staff

SASHA ALLISON, Research Associate *(as of June 2022)*
DANIEL BURGER, Study Director, Senior Program Manager
JENNIFER COHEN, Senior Program Officer
ROBERT GASIOR, Program Officer *(as of January 2023)*
JUAN SANDOVAL, Senior Program Assistant *(until June 2022)*
JESSICA SIMMS, Program Officer
EMILY TWIGG, Senior Program Officer *(until December 2022)*

Consultant

JAN SUMMERS, Copy editor

Reviewers

This Consensus Study Report was reviewed in draft form by individuals chosen for their diverse perspectives and technical expertise. The purpose of this independent review is to provide candid and critical comments that will assist the National Academies of Sciences, Engineering, and Medicine in making each published report as sound as possible and to ensure that it meets the institutional standards for quality, objectivity, evidence, and responsiveness to the study charge. The review comments and draft manuscript remain confidential to protect the integrity of the deliberative process.

We thank the following individuals for their review of this report:

ARUN AGRAWAL, University of Michigan
MARTHA ARRIETA, University of Southern Alabama
LACEY CAVANAUGH, Louisiana Department of Health
TIMOTHY COLLINS, University of Utah
ANN-MARGARET ESNARD, Georgia State University
ATYIA MARTIN, All Aces, Inc.
LAURA MYERS, University of Alabama
FRANKLIN NUTTER, Reinsurance Association of America (retired)
KEVIN SMILEY, Louisiana State University
SHANNON VAN ZANDT, Texas A&M University

Although the reviewers listed above provided many constructive comments and suggestions, they were not asked to endorse the conclusions or recommendations of this report nor did they see the final draft before its release. The review of this report was overseen by **SUSAN CUTTER,** University of South Carolina, and **DOUGLAS MASSEY,** Princeton University. They were responsible for making certain that an independent examination of this report was carried out in accordance with the standards of the National Academies and that all review comments were carefully considered. Responsibility for the final content rests entirely with the authoring committee and the National Academies.

Contents

PREFACE   xiii

SUMMARY   1

1   INTRODUCTION   17
     Study Context and Charge, 18
     Study Approach, 20
     Conceptualizing Compounding Disasters, 24
     Context and Framework for Compounding Disasters, 28
     Relevant National Academies Reports, 37
     Organization of the Report, 42

2   HAZARDS, EXPOSURE, VULNERABILITIES, AND
     DISASTER RISK IN THE GULF OF MEXICO REGION   43
     Hazards, 45
     Exposure, 55
     Vulnerability, 63
     Summary of Key Findings, 100

3   COMPOUNDING DISASTERS IN THE GULF OF
MEXICO REGION: 2020–2021   103
Synopsis of Disruptive Events, 103
Impacts of Compounding Disasters in Regions of Texas, Louisiana,
   and Alabama, 118
Harris and Galveston Counties, Texas, 121
Cameron and Calcasieu Parishes, Louisiana, 128
Baldwin and Mobile Counties, Alabama, 133
Key Themes from the Information-Gathering Sessions, 137
Summary of Key Findings, 147

4   INTERDEPENDENT SYSTEMS IN THE CONTEXT
OF COMPOUNDING DISASTERS   149
A Systems Approach to Understanding Compounding
   Disasters, 149
The Pandemic, Vulnerabilities, and System Interdependencies, 151
Impact of Regionality on Disaster Recovery, 164
Summary of Key Findings, 166

5   LESSONS LEARNED   169
Lessons Recognized and Learned, 170
Lessons Recognized and Learned: Themes from the
   Information-Gathering Sessions, 173
Lessons Recognized and Learned: NGOs, CBOs, and
   Government Officials, 184
Lessons Implemented, 185
Lessons Lost, 194
Summary of Key Findings, 196

6   CONCLUSIONS: REDUCING COMPOUNDING
DISASTER RISK BY ADDRESSING VULNERABILITIES
AND EXPOSURE AND BUILDING ADAPTIVE
CAPACITIES   199

REFERENCES   209

APPENDIXES

A   Glossary   247
B   Public Session Meeting Agendas   255
C   Commissioned Paper Abstracts   279
D   Committee Member and Staff Biographical Sketches   281

Preface

Disasters interrupt human life. Experiencing a single disaster—be it a hurricane, tropical cyclone, tornado, flood, severe winter storm, or global pandemic—can set off a domino effect of suffering and displacement from normalcy, home, and community for weeks, months, years, or even permanently. Throughout human history, these disruptions have reshaped communities and the prospects of current and future next generations to live healthy, fulfilling lives.

However, individual disasters do not occur in isolation within a community's history. So, what happens when multiple disasters coincide in time and space or occur in such rapid succession that a community has not recovered basic functionality before a new disaster recovery process begins? With each successive punch, vulnerabilities increase, heightening the risk that the next event could provide a knockout blow. These are compounding disasters.

The National Academies of Sciences, Engineering, and Medicine's Gulf Research Program commissioned this study to shed light on the particular experience of Gulf of Mexico communities that endured multiple disasters within an all too brief time frame. The cavalcade of disasters that struck the region between 2020 and 2021 afforded little opportunity to recover between events, stressed systems that underpin critical community functions, and brought an acute focus to the physical, health, and social vulnerabilities that predispose the region to potentially calamitous outcomes.

Compounding disasters extend and occur well beyond the Gulf of Mexico, yet the factors in play throughout the Gulf region offer a window—even a magnifying glass—to understand the complexities of risk and to tease out deep learnings and discover pathways that are broadly applicable to building adaptive capacity beyond the occurrence of individual events and most common hazards. While the inflection point of a weather or health disaster may be placed within a specific time frame, the effects often persist for years, resulting in the destabilization of communities across generations.

Why this report now? Most in public policy and disaster management take too limited a view of disasters—focused on the event directly in front of them. Given the interconnected realties of disasters that transcend temporal and spatial boundaries, a different view is required.

As a nation, it is imperative and increasingly urgent that we more fully communicate the complex nature of the threats that confront our communities. Rapidly intensifying and stronger tropical cyclones may increase due to climate change. We do not choose the time or place of the next pandemic. By continuing to build more in harm's way and failing to address societal vulnerabilities, we are changing the baseline denominator of disaster risk. Even if the storms were comparable in intensity to prior eras, the effect of those storms continues to grow (sometimes exponentially) as we expand the bull's eye with more structures and more people in the pathway.

When more people are within the zone of impact and then the frequency or severity ramps up, we are on course for more hardship and displacement. Without learning the lessons from our past and growing the adaptive capacities of communities, the effects of compounding disasters will overwhelm us.

This report would not have been possible without the generous contribution of numerous community stakeholders, leaders, and experts. The Committee on Compounding Disasters in Gulf Coast Communities, 2020–2021: Impacts, Findings, and Lessons Learned is particularly grateful to the community leaders from Houston-Galveston, Cameron-Calcasieu, and Mobile-Baldwin, whose personal experience and understanding of these issues is unrivaled and coupled with an unwavering commitment to lean in and make necessary changes before the next disaster occurs in their communities. Additionally, the committee thanks Dr. Stephen Strader, Villanova University, and Dr. Jennifer Trivedi, University of Delaware, for their scholarship and production of commissioned papers that augmented the committee's fact-finding process. Dan Wegendt provided his creative talents in helping the committee graphically depict compounding disaster risk. Finally, the committee acknowledges the staff from the National Academies of Sciences, Engineering, and Medicine who served this committee with distinction. Complex concepts, intricate logistics, and insightful editing came together to create a fulsome report and conclusions because of their tireless investment and support.

Roy E. Wright, *Chair*
Committee on Compounding Disasters
in Gulf Coast Communities, 2020–2021:
Impacts, Findings, and Lessons Learned

Summary

COMPOUNDING DISASTERS IN
THE GULF OF MEXICO REGION, 2020–2021

During 2020–2021, two global phenomena—the COVID-19 pandemic and climate change—shaped disaster risks worldwide, with particularly acute consequences in the Gulf of Mexico (GOM) region. The emergence and spread of COVID-19 transformed the public health risk of all other disaster events by amplifying health-compromising exposures and underlying vulnerabilities in the region. The pandemic also modified preparedness, response, and recovery procedures for the 2020–2021 extreme weather-climate events. The combined result was a complex and unprecedented public health and socioeconomic crisis—an exemplar of compounding disasters.

When society experiences a disaster, the impacts are typically attributed to an individual event, and damages stemming from disasters are quantified in terms of economic losses and direct fatalities. The National Oceanic and Atmospheric Administration's application of this conventional approach designated each of the seven major hurricanes and winter storm in the GOM in 2020 and 2021 as a billion-dollar disaster, and the COVID-19 pandemic has been characterized by some as a trillion-dollar disaster. These monikers, however accurate, do not reflect the human toll and disparate effects caused by multiple events that increase underlying physical and social vulnerabilities, strain adaptive capacities, and ultimately make communities more sensitive and likely to experience future disruptive events as disasters. Such is the case for many GOM communities that were still in varying states of recovery from previous disasters when they were impacted by the compounding disasters of 2020–2021.

As climate change continues to affect every region on Earth in a multitude of ways, the adverse impacts on communities will only increase. Humans and the environments they inhabit will experience more extreme temperatures; stronger storms; more intense rainfall; and rising, warmer seas. These circumstances are influencing the chaotic characteristics of extreme weather-climate events, including the intensity, duration, and regional impact of occurrences.

In the GOM region, many communities are experiencing the increasing risk of the consequences of compounding disasters as the result of combined hazard exposure and vulnerabilities, many rooted in historic systemic and structural racial discrimination, including underinvestment in infrastructure and housing, persistent poverty, and land-use planning decisions that have marginalized people and placed them in harm's way. These circumstances constrain the ability of residents to recover fully from disasters while increasing their sensitivity to the escalating effects of climate change and extreme weather-climate events. Consequently, perpetual disaster recovery, coupled with increasing disaster risk, is an enduring reality for many living in GOM communities.

STUDY CHARGE AND SCOPE

In 2022, the Gulf Research Program of the National Academies of Sciences, Engineering, and Medicine commissioned this study and tasked a seven-member ad hoc Committee on Compounding Disasters in Gulf Coast Communities, 2020–2021 with examining the unique characteristics and effects of the 2020–2021 compounding disasters in the GOM region, and examining how to manage and minimize the effects of these disasters on those who live and work in the region. Committee members were invited to serve because of their extensive academic and professional expertise in community resilience; disaster management; public health; and behavioral, social, and atmospheric sciences. The committee's charge included the following tasks:

- Describe the major disasters that occurred in the Gulf region during the 2-year period and, to the extent known, their impacts on community markers such as local economies, government functions, industry, the education system, human health, and social structure.

- Discuss how these impacts were compounded by the sequencing of events, as well as the factors that may have amplified these impacts (e.g., poverty, health disparities, economic and governance constraints, obsolete or inadequate infrastructure).
- Identify lessons learned from and needs exposed in how communities dealt with these compounding disasters.
- Provide findings on what factors enabled or could enable communities to successfully plan for, respond to, recover from, and mitigate the impacts of compounding disasters.

With nascent research on this topic, particularly within the context of the GOM during the period 2020–2021, the committee emphasized qualitative, primary data collection through information-gathering sessions with key informants in the GOM region and the review of locally generated after-action reports and other primary and secondary sources as its approach to the statement of task. Additionally, the committee commissioned two papers to supplement its deliberations: the first provides an exploration of health and community baseline conditions and disaster impacts in two Mississippi counties; the second offers a geospatial analysis of localized hazards, vulnerability, and exposure across the GOM region. These learnings were then contextualized within relevant scientific literature, including completed National Academies consensus study reports, to support the development of the committee's findings and conclusions. This methodology allowed the committee to meet its task by documenting compounding disasters' effects in a specific region within a specific time frame, from the lived experience and perspectives of those affected. The committee was not tasked with offering recommendations.

CONCEPTUALIZING COMPOUNDING DISASTERS

Drawing on existing frameworks derived from disaster scholarship, the committee began its inquiry by examining a conventional Venn diagram that contemplates disaster risk as the product of intersecting hazards, exposure, and vulnerability, as illustrated by Figure S-1.

Within each lens of this framework lie numerous layered and interrelated drivers of risk. For the purposes of this report, the committee adopted the use of the following definitions:

FIGURE S-1 Disaster risk as a product of intersecting hazards, exposure, and vulnerability.
SOURCE: Adapted from Figure 19-1 in M. Oppenheimer, M. Campos, R. Warren, J. Birkmann, G. Luber, B. O'Neill, and K. Takahashi, 2014: "Emergent risks and key vulnerabilities"; in *Climate Change 2014: Impacts, Adaptation, and Vulnerability. Part A: Global and Sectoral Aspects. Contribution of Working Group II to the Fifth Assessment Report of the Intergovernmental Panel on Climate Change*, C. B. Field, V. R. Barros, D. J. Dokken, K. J. Mach, M. D. Mastrandrea, T. E. Bilir, M. Chatterjee, K. L. Ebi, Y. O. Estrada, R. C. Genova, B. Girma, E. S. Kissel, A. N. Levy, S. MacCracken, P. R. Mastrandrea, and L. L. White, eds. Cambridge University Press, United Kingdom and New York, pp. 1039–1099.

- *Hazards* are understood to be processes, phenomena, or activities with the potential to "cause loss of life, injury or other health impacts, as well as damage and loss to property, infrastructure, livelihoods, service provision, ecosystems, and environmental resources" (USGCRP, 2023). Hazards may be natural or human-made and can occur individually or simultaneously. Examples of hazards reflected in this study include severe and extreme weather events and the global COVID-19 pandemic.
- *Exposure* describes the presence of people, infrastructure, housing, production capacities, and other tangible human assets in hazard-prone geographies or situations.
- *Vulnerability* encompasses social and economic sensitivities of individuals and groups, along with deficiencies in the structures and

systems on which they rely, that reduce their capacity to withstand hazards. Vulnerable communities are those least able to anticipate, cope with, and recover from these disruptive events.

- *Disaster risk* is expressed as a product of hazard, exposure, and vulnerability variables and understood as the potential for loss of life, injury, physical damage, or destruction resulting from the occurrence of one or more disruptive events in a given period.

Accordingly, realized disaster impacts are products of not only the severity or intensity of the hazard itself, but also the combination of vulnerability and exposure—or *sensitivity*—of the community and its underpinning systems and functions to suffer loss and damage. Many communities in the GOM region exist in a state of heightened disaster risk and perpetual recovery. As a result, the systems and functions that underpin communities are progressively diminished. From a humanitarian perspective, the immediate, visible, and experiential effects of disasters cannot be decoupled from the preexisting conditions of exposure and vulnerability that produce sensitivity to the event while also increasing sensitivity and risk to future events. As such, the disaster management enterprise must be responsive not only to the occurrence of singular and multiple disruptive events but also to the persistent societal conditions that create sensitivity to future disasters and compose the epoch nature of the compounding effects of disasters within disaster survivors' lives.

To guide its exploration of this topic, and for the purposes of this study, the committee characterizes a compounding disaster as **the result of overlapping, concurrent, or successive disruptive events that affect the societal, governmental, and/or environmental functions of a community or region and diminish the community's capacity to recover and resume essential activities. The weakening of these interrelated functions inhibits and prolongs the disaster recovery period, making communities more likely to experience amplified negative effects of future disruptive events. Some communities are at disproportionate risk of suffering the effects of compounding disasters as a result of the interplay of persistent physical and social vulnerability factors and increased exposure to climatic and non-climatic hazards.**

There is nothing linear about the movement of hazards, exposure, vulnerability, and the resulting risk. Through its deliberations, the committee found the conventional Venn diagram and its various two-dimensional iterations constrained in representing the multidimensional dynamics of compounding disasters. The common definitions of hazard(s), exposure, vulnerability, and risk continue to ring true, yet the ways that risk compounds requires a more integrative visualization, as depicted in Figure S-2.

The interaction and agitation within and among these variables is dynamic and complex. Realized hazards, experienced by communities as disruptive events, stack upon each other and exacerbate vulnerability and exposure. As the impacts of disruptive events compound, they push the interconnected societal mesh spanning multivariable exposure and vulnerability planes toward ever-increasing risk to future disruptive events.

Exposure and vulnerability increase (and decrease) by a range of choices made affecting variables within a community. These choices affect the adaptive capacity of a community to prepare for, respond to, and recover from disruptive events and disasters. As described throughout the report, increasing adaptive capacity is an intentional and effective way to pull down risk by reducing variables associated with exposure and vulnerability within and across societal systems.

FEATURES OF COMPOUNDING DISASTERS

Among the features of compounding disasters, four stand out as especially salient to GOM communities and the purpose of this study: (1) the disproportionate risk faced by vulnerable communities; (2) outcomes that result when that risk becomes realized impact; (3) effects

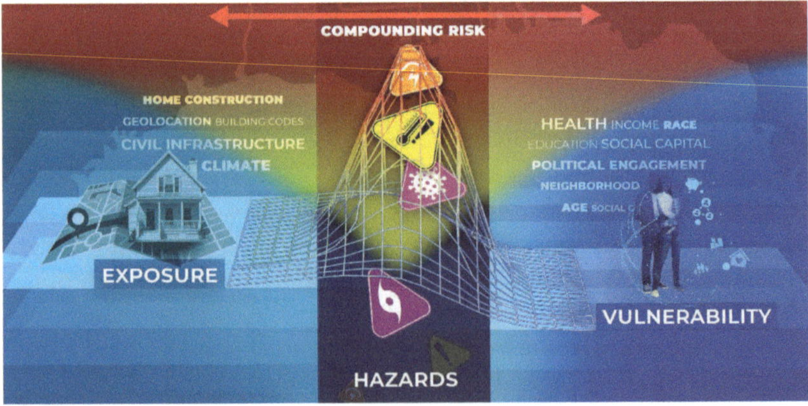

FIGURE S-2 Conceptual illustration of compounding disaster risk.
SOURCE: Image provided courtesy Insurance Institute for Building and Home Safety (IBHS).

on the interdependent systems and functions that serve both the routine and the acute needs of a community; and (4) overlapping, interrupted, and prolonged disaster recovery.

Particular Risk to Vulnerable Communities

Disasters manifest when a community's capacity to prepare for, respond to, and recover from the occurrence of a disruptive event is inadequate and overwhelmed. When compounding effects occur over time communities may bear the hardship of disaster without ever reaching traditional thresholds for a formal disaster declaration. It is the *sensitivity* of those who experience the brunt of the disruptive event that dictates whether the experience crosses the threshold to become a disaster. Heightened socioeconomic vulnerabilities and disaster-weakened systems and functions make many GOM communities particularly sensitive to future disruptive events and their compounding effects, regardless of the events' intensity. Many frontline communities (those who are highly exposed to climate risks due to where they live and the projected changes expected to occur in those places) have fewer resources; lower adaptive capacity; weak social or economic safety nets; and/or are underrepresented in policy, governance, and recovery planning (USGCRP, 2023) to be able to reduce sensitivity.

The committee explored the evidence-based disaster risk profile of GOM communities, including first-hand accounts of information-gathering session participants about the pandemic and extreme weather-climate events' impacts in 2020–2021. The events discussed were not restricted to federally declared or billion-dollar disasters, but included "smaller" disruptive events (e.g., heavy precipitation events, isolated tornados) that can compound in their effects and produce large-scale impacts. Risk profiles include components of physical, health, and social vulnerability, and exposure to extreme weather-climate events and hazards familiar to the region.

When Disaster Risk Becomes Realized Impact

The time frame for this study provides insight into outcomes that result when disaster risk becomes repeated realized impacts. Through a series of facilitated regional dialogues, detailed in Chapter 3, the committee engaged community leaders and practitioners who experienced multiple disruptive events in 2020–2021. Although the type, intensity, and location of these events varied, their rapid succession resulted in shared experiences across regions. Throughout its process, the committee sought to better understand the realized impact of multiple disruptive events on the function, resilience, and well-being of disaster-affected communities, including those who work in service to them.

Effects on Interdependent Community Systems and Functions

Chapter 4 details the widespread and profound compounding disaster impacts on the interdependent systems and functions that enable communities to function properly, and often result in prolonged—if not indefinite—recovery. Moreover, the longer recovery periods extend, for any household or community, the more likely it is that damage, losses, and the human toll will be compounded by one or more future disruptive events. In Southwest Louisiana, the February 2021 winter storm brought extreme cold weather with freezing rain, causing pipes to burst and acutely affecting water systems for homes and public buildings such as hospitals and dialysis centers that were still in disrepair from a previous pair of hurricanes. Three months after the winter storm, a heavy May rainstorm brought a foot of rain in less than 24 hours to areas affected by Hurricanes Laura and Delta in 2020, further exacerbating ongoing recovery efforts. Debris not yet removed from the 2020 storm sequence clogged stormwater systems, limiting the capacity available to absorb or drain the record rainfall. The floodwaters initiated yet another round of losses and claims into an already overwhelmed claims and assistance system that was unable to quickly mobilize funds into Southwest Louisiana, particularly Lake Charles.

Overlapping, Interrupted, and Prolonged Recovery

In many cases, the rapidity of event occurrence, coupled with overlapping, interrupted, and prolonged recovery periods, made it difficult if not impossible for residents, municipalities, claims adjusters, and responding federal agencies to differentiate among or ascribe realized impacts to specific disruptive events. Chapters 3 and 4 provide examples, including how in both Alabama and Louisiana, two hurricanes made landfall in such rapid succession that residents, building officials, and claims adjusters had difficulty accurately assigning damage to the causal event. As a result, many residents experienced delays or were unsuccessful in their attempts to effectively file claims and receive reimbursements or public assistance for disaster damages. As the disaster recovery period is prolonged, sensitivity to the occurrence of future disruptive events is heightened.

Lessons Learned and Lessons Lost

The compounding disasters within the time frame of this study cannot be isolated from prior disasters. Each region's session participants acknowledged being in some state of recovery from the physical and socioeconomic impacts of previous disasters, such as Hurricanes Katrina and Rita (in 2005), the *Deepwater Horizon* disaster (in 2010), Hurricanes Gustav and Ike (in 2008), and Hurricane Harvey (in 2017). Despite their widespread and devastating impacts, these prior events provided policy- and decision-makers with opportunities to learn from experience and make improvements to reduce future disaster risk. Chapter 5 examines how these experiences from previous disasters influenced communities' experiences in 2020–2021, while identifying barriers that may inhibit the uptake of lessons learned.

CONCLUSIONS:
REDUCING COMPOUNDING DISASTER RISK BY ADDRESSING VULNERABILITIES AND EXPOSURE AND BUILDING ADAPTIVE CAPACITIES

The committee's information-gathering, analysis, and deliberations yielded a number of conclusions regarding the compounding disasters that occurred in the GOM region in the 2020–2021 time frame. Grounded in the profound experiences relayed to the committee by the affected communities during the course of this study and reinforced with the scientific evidence base, the conclusions presented here and in Chapter 6 underscore the need to reimagine efforts that support disaster preparedness, mitigation, and recovery in an era of intensifying extreme weather-climate events and increased risk of compounding disasters. Conclusions 1–3 address the expanding realities of compounding disaster risk; conclusions 4–6 identify the need for improving and expanding our understanding of the new scale and temporal scope of compounding disasters; conclusions 7–9 refer to the need to bolster public health, mental health, and community-based organizations and their adaptive capacity; and conclusions 10–14 identify new approaches to contending with compounding disasters. The conclusions are based on the committee's information gathering and analysis of disasters in

GOM communities, and they are not necessarily generalizable to communities experiencing disasters in other regions.

The Expanding Realities of Compounding Disaster Risk

Conclusion 1: Compounding disasters introduce new, interconnected, and complex risk scenarios due to the increased potential of multiple hazards overlapping in time and space.

As established in Chapter 2, recent studies in attribution science show that climate change is causing an increase in the frequency and/or severity of tropical storms, heavy rainfall, and extreme temperatures. The intensification of these and other more extreme weather-climate events are intersecting with areas experiencing high levels of health disparities, social vulnerabilities, and increased exposure due to population growth in hazard-prone areas. This convergence is resulting in prolonged and overlapping periods of disaster recovery. While hazards may be unalterable, their impact can be reduced by collectively building adaptive capacity—whether preemptive, in real time, or through the recovery process—within vulnerable and exposed systems, institutions, and people.

Conclusion 2: Increased compounding disaster risk requires communities to plan and prepare for the co-occurrence of multiple and varied disruptive events that interact with societal exposure and vulnerabilities to amplify overall disaster impact.

Two phenomena of global scope and scale—the sudden emergence and ongoing evolution of the COVID-19 pandemic and climate change—fundamentally influenced concurrent disaster events more circumscribed in time and place. The evidence presented in Chapter 3 makes clear that while GOM communities have plans in place for addressing more common climate hazards (i.e., hurricanes), the same level of planning did not exist for less common events (i.e., Winter Storm Uri, the COVID-19 pandemic), let alone the co-occurrence of multiple events. Advancing adaptive capacity involves understanding how disaster events impact specific localities and funding locally led, long-term planning for risk reduction and climate adaptation programs. Such programs must advance with built-in flexibility to respond to evolving priorities.

Conclusion 3: When a community increases its capacity to absorb the effects of hazards and minimizes its recovery needs, disaster effects are less likely to compound.

Targeted, community-guided investments to increase the resilience of essential services and infrastructure are important to achieve this objective. While it is preferable to invest before a disaster occurs, for those communities able to access such investments, the influx of recovery funding after a disaster can provide an opportunity to mitigate future losses and the potential for compounding disasters. As discussed in Chapter 5, many information-gathering session participants highlighted the importance of flexible pandemic relief funds as an invaluable asset to help their communities recover. The literature on social capital discussed in Chapter 2 and the comments made by information-gathering session panelists both emphasize the critical role of social capital and cohesion in building resilience and catalyzing recovery from disaster events. Investments into building physical capacity to withstand hazards are essential, but the strengthening of adaptive capacity, including formal and informal relationships among community stakeholders is equally foundational to disaster resilience.

The New Scale and Temporal Scope of Compounding Disasters

Conclusion 4: Perception and understanding of risk are commonly grounded in past experience, leading to complacency in preparation and mitigation.

As discussed in Chapter 3, cognitive biases such as recency bias and normalcy bias can lead to inaccurate conceptualizations of past events and can hamper risk communication, mitigation, and planning for future events that extend beyond what has been experienced or is perceived to be the benchmark extreme. These biases are similarly reflected in emergency management protocols, land-use planning and plans, zoning regulations, public utility design, and building codes, which are often grounded in historical precedent or probabilistic hazard descriptions derived from historical data. Given a changing climate, this hindcasted vantage is unlikely to be representative of future hazard risks. Overcoming these biases requires new strategies.

Conclusion 5: Effective disaster recovery requires an "epoch" rather than "event" view that more fully captures the prolonged effects of compounding disasters and reflects the experienced reality of the community.

In the information-gathering sessions summarized in Chapter 3, participants continually reported that the impacts of discrete disaster events were impossible to disentangle when the disasters occurred in such rapid succession and their recovery time lines overlapped. This was particularly true in socially vulnerable communities that were inadequately resourced before disaster events. An event-based perspective on disaster management is inherently narrow, reactive, and artificially time constrained. The event-driven view focuses on the symptoms rather than the root causes of disaster losses. Shifting to an epoch view better frames the breadth of lived experiences with disasters and accounts for the potential for compounding losses, driving the broad disaster response and recovery enterprise to more comprehensive and effective pathways forward.

Conclusion 6: Risk assessment and communication for extreme weather-climate and multiple/sequential events is inadequate for both current and future conditions.

The severity of many of the disaster impacts in 2020–2021 exceeded expectations. With a changing climate, the potential for complex events only increases. As described in Chapter 3, many information-gathering session participants brought to light a variety of disaster communication challenges that were exacerbated by the compounded nature of events, including inconsistent technology and broadband access, accommodating non-literate populations, lack of public trust in government, and the spread of misinformation. Better preparation for future compounding events requires incorporation of compound event risk assessments, multisector collaboration, and improved risk communication. The multiple levels of a participatory information-sharing approach will increase understanding of the unique needs of socially vulnerable groups, enhance transparency, and reduce the potential for misinformation.

Bolstering Adaptive Capacity

Conclusion 7: Health care and public health systems will require increased adaptive capacity and staffing to respond to diverse challenges posed by compounding disasters.

Compounding disasters notwithstanding, health systems in the GOM region are already vulnerable and struggle to meet the needs of GOM populations, as discussed in Chapter 2. Recent literature and information-gathering session participants made clear the increased strain on health systems caused by the co-occurrence of natural hazards and the COVID-19 pandemic. The nature of the hazards encountered influences the impacts on population health, health systems, and public health services. Even as extreme weather-climate events damage facilities and disrupt access to health care, emerging disease outbreaks require physically intact facilities and enough frontline professionals staffing them so that services can be rapidly shifted to communicable disease care, and local public health capacity can be maintained.

Conclusion 8: Pervasive mental health impacts of compounding disasters undermine the adaptive capacity of communities to withstand and effectively recover from disruptive events.

Information-gathering session panelists spoke extensively about the negative mental health impacts of compounding disasters on survivors, especially for specific subpopulations, such as first responders and volunteers. The impacts of disasters on behavioral health are pervasive, extend across a spectrum of severity, and may continue for a prolonged time. Chapter 2 details some of these specific impacts and explores the links between disaster exposure and adverse psychological and behavioral health outcomes. Mental health needs are magnified for those who experience more intense exposure and vulnerability to hazards. Particular attention is necessary to the mental health needs of those who are disproportionately affected psychologically, including children; the elderly; medically high-risk patients, including those with severe mental illness; and the frontline professionals including first responders, public health professionals, and volunteers tasked with disaster response and recovery responsibilities.

Conclusion 9: The heavy dependence on community-based organizations can strain these individuals and groups beyond the point of effectiveness in the face of compounding disasters.

In the aftermath of a disaster, many immediate needs are met by neighbors helping neighbors and community-based organizations. As discussed in Chapter 4, information-gathering session participants, particularly those representing nongovernmental and community-based organizations, consistently reported feeling burned out and over-whelmed from spending long hours providing physically and/or emotionally taxing labor for little compensation. While these volunteers and organizations are essential to disaster response and recovery, they may suffer from the compounded disaster impacts themselves and be unable to assist when needed most. Overreliance on this often uncompensated or underpaid workforce risks the depletion of critical human resource capacity for effective long-term recovery from successive disasters.

New Approaches to Contending with Compounding Disasters

Conclusion 10: While a powerful tool for delivering services in times of crisis, technology is not a universal substitute for interpersonal communication and in-person disaster recovery assistance.

COVID-19 restrictions forced rapid innovations, creating more efficient ways to share critical data digitally and greater agility in delivering services virtually—adaptive shifts that were invaluable to service continuity when storms disrupted physical operations. However, as discussed in Chapter 4, processing federal disaster recovery assistance requests and insurance claims virtually caused errors and inefficiencies for survivors. Vulnerable populations, notably the elderly and low socioeconomic status individuals, are often least equipped to navigate complex and continually evolving online systems. For these populations, the reduction or elimination of in-person assistance prolonged disaster recovery efforts and increased risk for future disruptive events.

Conclusion 11: Safe, sanitary, and secure housing is a fundamental determinant of disaster resilience and recovery, and can thus be viewed as core community infrastructure.

As established in Chapter 2, access to stable, long-term housing is critical for individual and collective well-being, and displacement from secure housing has deleterious impacts on security, mental health, social connectedness, and well-being broadly. Since the most vulnerable housing (older/deferred-maintenance properties) is often occupied by the most vulnerable households, this segment of the inventory is especially susceptible to compounding disasters, even as the result of lower-intensity weather-climate events. Areas experiencing economic downturns or with aging housing inventories require particular attention to housing quality as part of ongoing efforts to address the broader affordable housing crisis that existed well before 2020.

Conclusion 12: Effective community-guided risk reduction for compounding disasters requires greater understanding of and planning for the full range of potential disruptive events, along with their cumulative effects.
Chapter 2 outlines the wide range of both hazards to which GOM populations are exposed and the impacts that these hazards can have on human health and well-being. These can be low- or high-probability events, and can span climatic (e.g., rapidly intensifying storms, slow-moving and stalling storms dropping significant rainfall, temperature extremes) and non-climatic (e.g., pandemic, technological) scenarios, including worst-case extremes. Local-level participatory planning processes that routinely engage and collaborate with socially vulnerable communities that are marginalized by lack of representation in policy, governance, and recovery planning; income; education; age; ethnicity/race; gender; sexual identity; and/or medical risk and that are disproportionately affected by compounding disasters will more effectively guide and prioritize efforts to reduce potential impacts and concurrently build adaptive capacity.

Conclusion 13: Stronger mechanisms are essential to translate lessons recognized from prior experience into lessons learned and implemented.
As discussed in Chapter 5, lessons are often identified after disasters, but they are rarely codified into formal policies and procedures before the next disaster strikes. Interaction between professionals and community members can relay local disaster memories to agency personnel. Ongoing and inclusive education, training, drills, and scenario-based exercises can perpetuate and reinforce lessons learned among all affected communities and decision-makers. Collaborative input of lessons learned into after-action reports can increase knowledge base, participation, and potential for implementation. Efforts at all levels of government to dissolve institutional silos and reinforce improved practices can also sustain lessons learned between disasters.

Conclusion 14: Revisions to disaster planning, response, and recovery policies and procedures need to directly address and eliminate the uneven access to resources that can exacerbate social and economic inequities in the wake of disasters.

Residents in affected areas with the most financial means, social capital, and technological skills are able to more effectively access recovery resources; compounding disasters can leave those without such critical capabilities in a more distressed condition, and even minor disruptive events can become disasters. As discussed in Chapter 2, a large portion of GOM residents are in some capacity marginalized, and the current disaster recovery process often exacerbates vulnerabilities rather than addressing these vulnerabilities at their root. Information-gathering session participants highlighted the need to incorporate equity into the disaster recovery process to better assist marginalized, socially vulnerable community members.

The climate-intensified stress on people, communities, and the systems and functions that support them; disruptive events; and disaster risk will likely continue to increase, along with the likelihood that more communities will experience the debilitating effects of compounding disasters. The committee recognizes that transdisciplinary research and applied science in disaster resilience continues to advance in the wake of major events in the United States and around the world. This study is intended as a contribution not only to the evidence base but in support of continued research and calls for collective community-guided action and investment to reduce systemic vulnerabilities, mitigate hazard exposure, and increase adaptive capacities to enable communities to address effectively the complex challenges that confront them.

1

Introduction

Experiencing a disaster—a hurricane, tornado, flood, severe winter storm, or global pandemic—wreaks havoc on the lives and livelihoods of individuals, families, communities,[1] and entire regions. For people who live in the five U.S. states outlining the coastline of the Gulf of Mexico (GOM) region—Texas, Louisiana, Mississippi, Alabama, and Florida—living through and trying to recover from such a disaster is not a once-in-a-lifetime experience but an enduring reality. Between 2020 and 2021, seven major hurricanes and a severe winter storm (later named Uri) severely affected communities across the GOM region, with each of these events ultimately being designated a billion-dollar disaster.[2] These storms, some arriving in the same region within weeks of one another, occurred while communities were coping with the illness, death, uncertainty, and disruption brought about by the COVID-19 pandemic. This constellation of sequential, sudden-onset disasters imposed compounding stress on a region with

---

[1] Community, as it is defined in this report, predominantly refers to "members of a collectivity, who share a common territorial area as their base of operation for daily activities." In some instances, it refers to a social group whose members are bound together by the sense of belonging created out of everyday contacts covering the entire range of human activities (Parsons, 1969, as cited in NASEM, 2021).

[2] The term "billion-dollar disaster" refers to weather and climate events for which the overall economic damages/costs reached or exceeded $1 billion (including Consumer Price Index adjustment to 2023). For more information see NOAA's National Centers for Environmental Information overview on Billion-Dollar Weather and Climate Disasters (NCEI, n.d.-a).

multifaceted and persistent socioeconomic and socioenvironmental disparities. As a result of these disparities, communities with heightened vulnerability and exposure to hazards experienced disproportionate impacts from
disaster events. To these communities, the moniker billion-dollar disaster
fails to encapsulate a broader scope of physical, economic, social, cultural,
and emotional damages caused by unrelenting disasters. These communities are on the frontlines[3] of climate change, which is exacerbating chronic
environmental stressors and driving the intensification of dangerous storms
and other extreme weather-climate events.[4] In this context, new strategies
are essential to enable communities to plan for the increasing risk associated
with compounding disasters.

STUDY CONTEXT AND CHARGE

The Gulf Research Program (GRP)[5] of the National Academies of
Sciences, Engineering, and Medicine (the National Academies) sponsored
this consensus study to explore factors that enable—or could enable—Gulf
Coast communities to prepare for, respond to, mitigate, and recover from
disasters more effectively. With the events and lived experiences of the
2-year period 2020–2021 to guide the study, the GRP appointed an ad
hoc committee of experts in community resilience; disaster management;
public health; and behavioral, social, and atmospheric sciences (Committee
on Compounding Disasters in Gulf Coast Communities, 2020–2021) and

---

[3] Frontline communities include "people who are both highly exposed to climate risks
(because of the places [in which] they live and the projected changes expected to occur in
those places) and have fewer resources, capacity, social or economic safety nets, or political
power to respond to those risks. Frontline communities are those that experience the 'first
and worst' consequences of climate change. These are often, but not limited to, communities
of color and low-income communities" (USGCRP, 2023).

[4] "Extreme weather-climate event" is a hybrid term used by the committee to describe
weather events that are more likely to occur, more intense, longer-lasting, or larger in scale due
to climate change. See Chapter 2, Attribution Science, and glossary definitions for Extreme
Event, Weather-Related, and Extreme Event, Climate-Related for details.

[5] The GRP was established in 2013 with funds from a criminal settlement stemming
from the *Deepwater Horizon* disaster, which occurred in the Gulf of Mexico in 2010. Upon
receipt of the settlement funds, the National Academies, through the GRP, was charged with
establishing a 30-year program in the public interest focused on offshore energy safety, environmental protection, and human health. In sponsoring this study, the GRP aims to advance
scientific understanding, engage networks, and ultimately bridge knowledge to action in
accordance with its primary strategic goals.

asked them to provide findings and conclusions relevant to their task (see Box 1-1 for the committee's statement of task).

In an era of rapidly increasing climate-amplified hazards and extreme weather-climate events, it is crucial for policymakers, decision-makers, and communities to understand how impacts associated with disruptive events and chronic stressors interact, exacerbate health disparities and socioeconomic stressors, and inhibit effective or complete disaster recovery.

BOX 1-1
Statement of Task

A committee of the National Academies of Sciences, Engineering, and Medicine will document the impact and lessons learned from multiple, compounding disasters that occurred across the Gulf Coast region in 2020–2021. Within this 2-year period, Gulf of Mexico communities experienced numerous billion-dollar extreme weather disasters, widespread energy infrastructure failures, and the COVID-19 pandemic. These events occurred against a backdrop of persistent social and economic inequities, resulting in disproportionate impacts of these events on historically marginalized, disadvantaged, and excluded groups. The increasing frequency or severity of extreme weather events, amplified by climate change, will likely continue to exacerbate these impacts on the resilience of communities throughout the Gulf region.

This committee will

- Describe the major disasters that occurred in the Gulf region during the 2-year period and, to the extent known, their impacts on community markers such as local economies, government functions, industry, the education system, human health, and social structure.
- Discuss how these effects were compounded due to the sequencing of events as well as the factors that may have amplified these impacts (e.g., poverty, health disparities, economic and governance constraints, obsolete or inadequate infrastructure).
- Identify lessons learned from and needs exposed in how communities dealt with compounding disasters.
- Provide findings on what factors enabled or could enable communities to successfully plan for, respond to, recover from, and mitigate the impacts of compounding disasters.

This principle is especially relevant for GOM communities that remain in various stages of recovery from previous disasters, including hurricanes and the lingering effects of the *Deepwater Horizon* disaster of 2010. These compounding disaster effects undermine the systems and functions of many GOM communities, making them more vulnerable to experiencing future disruptive events as disasters. Referenced throughout this report, the committee calls attention to the need for strengthening the *adaptive capacity* of vulnerable and exposed communities. Simply defined, adaptive capacity is the relative ability of systems, institutions, and humans to adjust to potential damage, to take advantage of opportunities, or to respond to consequences (USGCRP, 2023). Enhancing adaptive capacity includes a range of strategies and actions, explored in greater detail throughout this report.

It should be noted that, although this study focuses on a particular geographic region, it is the committee's hope and intention that its findings and conclusions will inform national and international dialogues on disaster risk reduction, with an emphasis on the imperative to improve daily living conditions in at-risk communities, particularly those with high exposure and vulnerability and low adaptive capacity.

STUDY APPROACH

The topic of compounding disaster impacts and recovery is complex, and many applied research questions regarding strategies, solutions, and governance remain outstanding (NASEM, 2022a). With nascent research on this topic, particularly within the context of the GOM during the 2020–2021 period, the committee emphasized qualitative, primary data collection through information-gathering sessions with key informants in the GOM and the review of locally generated after-action reports and other primary and secondary sources as its approach to the statement of task.

These learnings were then contextualized within relevant scientific literature, including completed National Academies consensus study reports, to support the development of the committee's findings and conclusions. This method allowed the committee to meet its task by applying the evidence base and documenting the effects of compounding disasters in a specific region within a specific time frame, from the lived experience, knowledge, and assessments of those affected. The committee was not tasked with offering recommendations.

The committee conducted information gathering from key informants and subject-matter experts for this study from January 2022 through April

2023. Deliberations among committee members continued throughout the information-gathering and report development process and concluded in spring 2024.

The committee's work was informed by 13 in-person public information-gathering sessions held over 6 days in three locations, four virtual public sessions, a review of relevant scientific literature, and two commissioned papers.

Virtual Information-Gathering Sessions

The committee convened four virtual information-gathering sessions between December 2022 and April 2023 to gain perspective on several topics, including housing and disaster-related displacement; meteorological, atmospheric, and oceanographic influences on extreme weather-climate events; state and municipal chief resilience officers; and experiences of disaster-affected community stakeholders outside of the in-person sessions. All virtual information-gathering sessions were public. Agenda and participant listings are included in Appendix B.

Regional, In-Person Information-Gathering Sessions

The committee's choice of regional locations in which to hold information-gathering sessions was based on the states where multiple disruptive events significantly affected community functioning during the 2020–2021 time frame. While recognizing that disruptive events and disasters occurred in many communities throughout the GOM region, the committee ultimately selected for its analysis of compounding disaster experiences three representative GOM areas to conduct information gathering.[6] Together, these selections reflect the committee's priorities in seeking a range of compounding disaster experiences across GOM states with varying demographics, capacities, and experiences contending with disasters.

In October 2022, the committee convened five in-person information-gathering sessions in Houston, Texas, with participants who work and/or reside in Harris or Galveston County. In December 2022, the committee convened four in-person public information-gathering sessions in Lake

---

[6] Counties in Mississippi and Florida were not selected for in-person visits because of the absence of both weather-climate and hazard events with widespread impacts that significantly affected community functioning in the 2020–2021 time frame. At times, data from these states are included in the report for contextual purposes.

Charles, Louisiana, with participants who work and/or reside in Calcasieu or Cameron Parish. In January and February 2023, the third and final set of four in-person public information-gathering sessions was held in Mobile, Alabama, with participants who work and/or reside in Mobile or Baldwin County. In-person sessions included representatives from state and local government; federal agencies; and faith-based, community-based, and nongovernmental organizations involved in various aspects of disaster risk management. These experts were invited to share their personal and professional experiences and perspectives on the impacts of and interconnections among disasters and on lessons learned, forgotten, and applied, as well as policy and practice considerations related to compounding disasters in GOM states between January 2020 and December 2021. All in-person information-gathering sessions were public. Agenda and participant listings are included in Appendix B. The primary data elicited from session participants in the information-gathering sessions formed the basis of the report's summary, analysis, findings, and conclusions.

Methodological Limitations to In-Person Information-Gathering Sessions

The committee aimed to hear from a diverse array of representatives involved in the weather-climate event and disaster space. This approach is not without limitations. Time, resources, location, availability, residency, the particular dynamics and composition of attendees for each session, comfort levels of participants in sharing their experiences and perspectives with the committee or among their professional peers, the number of panelists, and their loquaciousness and interest in particular topics all shaped the information gathered. Coordinating speakers for the information-gathering sessions also depended on representatives' availability. Panelists who participated in these sessions also still lived and worked in the GOM region, and their perspectives and experiences may differ from those who opted to leave or were displaced by the disasters. Despite these potential biases and limitations, the insight gained from these sessions formed the basis for the committee's understanding of what it was like to live in the GOM region during this tumultuous period and how recovery has since progressed.

Primary and Secondary Sources

At the outset of its deliberations, the committee worked in collaboration with the National Academies Research Center to develop and carry out a review of the relevant literature to support its work and provide additional scientific evidence to inform this report and the committee's findings and conclusions. The committee also conducted literature searches to identify contemporaneous news articles as well as peer-reviewed and gray literature related to compounding disasters and associated impacts in GOM states within the 2020–2021 time frame.

After-Action Reports

It is common for government bodies to assemble "after-action" or "action" reports in the wake of disasters. Responding agencies such as the Federal Emergency Management Agency (FEMA), their local counterparts, and the National Oceanic and Atmospheric Administration (NOAA) seek to identify both successes and failures in disaster preparations and responses. The U.S. Department of Housing and Urban Development requires recipients of disaster recovery grants to prepare action plans that specify how dollars will be spent and best practices for recovery. These reports provide systematic reviews of how government agencies responded to disasters and how they might learn from any inadequate responses. The committee conducted a systematic search for such reports and reviewed them for specific practices that reflected lessons learned during 2020–2021. This report discusses after-action reports produced by the state of Louisiana, the city of Houston, and the state of Alabama. Additionally, several information-gathering session participants from government agencies participated in the production of these reports.

Commissioned Papers

The committee commissioned two papers to supplement its information-gathering sessions (see Appendix C). The first paper, "Compounding Disasters in Gulf Coast Communities 2020–2021: Impacts, Findings, and Lessons Learned in Jefferson Davis and Marion Counties, Mississippi," produced by Dr. Jennifer Trivedi, examines the physical, social, and economic factors contributing to the realized impact of multiple disasters within Jefferson Davis and Marion Counties, Mississippi, where the committee did not conduct a site visit or hold in-person information-gathering sessions, but where social vulnerabilities are significant. This paper also seeks to identify physical, social, and economic factors that may reduce the impacts of future disasters in these two counties. The second paper, "How a Changing Societal

Landscape Is Shaping Gulf Coast Tropical Cyclone and Tornado Disasters," produced by Dr. Stephen Strader, addresses physical and social factors contributing to the realized impact of multiple disasters within the particular geographic areas of interest for this study: Houston and Galveston Counties, Texas; Cameron and Calcasieu Parishes, Louisiana; and Mobile and Baldwin Counties, Alabama. Jefferson Davis and Marion Counties, Mississippi, were also included in the analysis.

CONCEPTUALIZING COMPOUNDING DISASTERS

Drawing on existing frameworks derived from disaster scholarship, the committee began its inquiry by examining the conventional Venn diagram that contemplates disaster risk as the product of intersecting hazards, exposure, and vulnerability, as illustrated by Figure 1-1.

FIGURE 1-1 Disaster risk as a product of intersecting hazards, exposure, and vulnerability.
SOURCE: Adapted from Figure 19-1 in M. Oppenheimer, M. Campos, R. Warren, J. Birkmann, G. Luber, B. O'Neill, and K. Takahashi, 2014: "Emergent risks and key vulnerabilities"; in *Climate Change 2014: Impacts, Adaptation, and Vulnerability. Part A: Global and Sectoral Aspects. Contribution of Working Group II to the Fifth Assessment Report of the Intergovernmental Panel on Climate Change*, C. B. Field, V. R. Barros, D. J. Dokken, K. J. Mach, M. D. Mastrandrea, T. E. Bilir, M. Chatterjee, K. L. Ebi, Y. O. Estrada, R. C. Genova, B. Girma, E. S. Kissel, A. N. Levy, S. MacCracken, P. R. Mastrandrea, and L. L. White, eds. Cambridge University Press, United Kingdom and New York, pp. 1039–1099.

Within each of these lenses lie numerous layered and interrelated drivers of risk. For the purposes of this report, the committee adopted the use of the following definitions:

- *Hazards* are understood to be processes, phenomena, or activities with the potential to "cause loss of life, injury or other health impacts, as well as damage and loss to property, infrastructure, livelihoods, service provision, ecosystems, and environmental resources" (USGCRP, 2023). Hazards may be natural or human-made and can occur individually or simultaneously. Examples of hazards reflected in this study include severe and extreme weather-climate events and the global COVID-19 pandemic.
- *Vulnerability* encompasses social and economic sensitivities of individuals and groups, along with deficiencies in the structures and systems on which they rely, and that reduce their capacity to withstand hazards. Vulnerable communities are those least able to anticipate, cope with, and recover from disruptive events.
- *Exposure* describes the presence of people, infrastructure, housing, production capacities, and other tangible human assets in hazard-prone geographies or situations.
- *Disaster risk* is expressed as a product of hazard, exposure, and vulnerability variables and understood as the potential for loss of life, injury, physical damage or destruction resulting from the occurrence of one or more disruptive events in a given period.

Accordingly, realized disaster impacts are products of not only the severity or intensity of the hazard itself but also the combination of vulnerability and exposure—or *sensitivity*—of the community and its underpinning systems and functions to suffer loss and damage.

Many communities in the GOM region exist in a state of heightened disaster risk and perpetual recovery. As a result, the community's systems and functions are progressively diminished. From a humanitarian perspective (UNHCR, 2024), the immediate, visible, and experiential effects of disasters cannot be decoupled from the preexisting conditions of exposure and vulnerability that produce sensitivity to the event while increasing sensitivity and risk to future events. As such, the disaster management enterprise must be responsive to the occurrence of singular and multiple disruptive events but also of the persistent societal conditions that create sensitivity to future disasters and compose the epoch nature of the compounding effects of disasters within disaster survivors' lives.

To guide its exploration of this topic, and for the purposes of this study, the committee characterizes a compounding disaster as

> **the result of overlapping, concurrent, or successive disruptive events that affect the societal, governmental, and/or environmental functions of a community or region and diminish the community's capacity to recover and resume essential activities. The weakening of these interrelated functions inhibits and prolongs the disaster recovery period, making communities more likely to experience amplified negative effects of future disruptive events. Some communities are at disproportionate risk of suffering the effects of compounding disasters as a result of the interplay of persistent physical and social vulnerability factors and increased exposure to climatic and non-climatic hazards.**

There is nothing linear about the movement of hazards, exposure, vulnerability, and risk variables. Through its deliberations, the committee found the classic Venn diagram (depicting hazards, exposure, vulnerability, and resulting risk) constrained in its representation of the dynamics and variables associated with compounding disasters. The common definitions of hazard, exposure, vulnerability, and risk remain accurate, yet the ways that risk compounds and grows require a more integrative visualization, as depicted in Figure 1-2. The interaction and agitation within hazards, exposure, vulnerability variables, and the resulting risk is more dynamic and complex.

To help visualize this complex interaction, the committee developed a conceptual illustration, Figure 1-2.

The scenario illustrated in Figure 1-2 represents multiple hazards that are propelled into interconnected societal mesh. The mesh represents the interwoven dimensions of exposure and vulnerability that largely influence the lived experience within a community and its sensitivity to overlapping, concurrent, or successive disruptive events caused by one or more hazards. The planes, or platforms, on which vulnerability and exposure variables are layered, are pushed upward into a deepening red zone by the occurrence of hazards, resulting in increased compounding disaster risk and heightened sensitivity to the occurrence of future disruptive events.

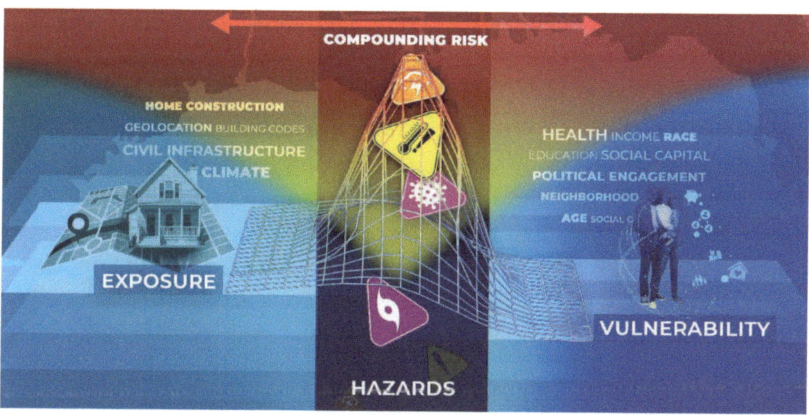

FIGURE 1-2 Conceptual illustration of compounding disaster risk.
SOURCE: Image provided courtesy Insurance Institute for Business and Home Safety.

In this example, the platforms for exposure and vulnerability are not equal, nor are the variables stacked upon them. In reality, the levels of exposure and vulnerability are unique to each community and dynamic in both space and time. Exposure to hazards increases or decreases through numerous variables, including land development patterns, housing realities, infrastructure location and functionality, and climate-driven changes within the physical environment. Vulnerability increases or decreases through socioeconomic strength and stability, health care and education access and quality, and social capital, among others. Exposure and vulnerability are inextricably linked and influenced by choices made by individuals, families, community leaders, and policymakers. Long-standing and systemic realities often dominate how exposure and vulnerability combine to create sensitivity within the context of disasters. As disasters compound, quality of life and the ability to prepare for and recover from future disruptions deteriorates. As described throughout the report, making choices to increase adaptive capacity within community systems may be the most effective way to pull down risk by intentionally reducing exposure and vulnerability.

Shared Compounding Disaster Experiences in the Gulf of Mexico Region

At the committee's information-gathering sessions, participants from the GOM region expressed views aligned with findings from numerous recent studies—that when disasters compound, resources become strained, and the preparedness, response, and recovery phases often overlap (Finucane et al., 2020; Knox et al., 2022; Mitchell and Knox, 2021; Singh et al., 2023). Simply, there is not enough time to respond to and recover from the first disruptive event by the time the second (or third) occurs, placing interdependent critical infrastructure systems and populations at further risk (Wells et al., 2022). As a result, decision-makers face difficult trade-offs when prioritizing recovery of needed services, such as infrastructure repair, debris cleanup, and system restoration (Knox et al., 2022). Compounding disasters also exacerbate and multiply systemic and socially constructed vulnerabilities that subject certain populations to disproportionate disaster impacts (NASEM, 2022a).

Participants in the Texas, Louisiana, and Alabama information-gathering sessions shared elements of three overarching and related experiences with compounding disasters: (1) material damage and losses compound, often in unanticipated ways; (2) recoveries interrupted result in extended recovery time lines, and many adverse impacts are compounded by the subsequent event(s); (3) high sensitivity and low adaptive capacity (e.g., low socioeconomic status and poor health outcomes) render individuals and communities more vulnerable to acute disaster impacts and compounded losses, and disadvantaged for long-term recovery.

The committee's engagement across the GOM region pointed to one overarching dynamic that transformed the lived experience of compounding disasters and living conditions within the communities—enhanced adaptive capacity. Referring back to Figure 1-2, risk can be lowered through the enhancement of adaptive capacity. When organized and activated, the people, systems, and institutions possess inordinate abilities to take advantage of opportunities to reduce vulnerabilities, lower exposures, and bend their response to disasters toward recovery and better outcomes.

CONTEXT AND FRAMEWORK FOR COMPOUNDING DISASTERS

While research on compound(ing) risks, events, and disasters is a rapidly growing field, many applied research questions regarding strategies, solutions, and governance remain outstanding (NASEM, 2022a). To inform some of these gaps from both multidisciplinary and empirical perspectives, the committee hopes to offer a more nuanced understanding of the realized effects of compounding disasters within a particular region and timescale. In addition, and as identified by the Fifth U.S. National Climate Assessment, released in November 2023, this report further explores the "intersection of climate hazards with other environmental hazards like . . . pandemics, or socioeconomic stressors like poverty and lack of adequate housing that disproportionately impact overburdened communities, thereby deepening existing societal inequities" and hindering the ability and capacity of such communities to "respond and cope" (Singh et al., 2023).

Because of climate change, evolving and dynamic weather patterns, and other emergent threats, the risks associated with disruptive events and disasters are also evolving. Drivers of climate change risk are creating new risks and exacerbating existing ones (Simpson et al., 2021). Disruptive events and disasters, especially if occurring in rapid succession, change the severity of risk and impact recovery in often unpredictable ways.

Disasters and disruptive events and their residual impacts are complex, interconnected, long-term phenomena that transcend temporal and spatial boundaries, cascading and compounding across sectors and regions (Lukasiewicz and O'Donnell, 2022). The compounding disasters of 2020–2021 in the GOM region created conditions that modified, via a compounding process, the impacts of the disasters that followed. For example, the impacts of Hurricane Laura, which made landfall on August 27, 2020, in Cameron Parish, Louisiana, were and continue to be disruptive, affecting a population and region already grappling with longer-term stressors such as inadequate housing, high unemployment, economic shocks, insufficient health care staff, and a pandemic. The hurricane winds tore off roofs, smashed windows, ripped out power lines and poles, flooded buildings, and overall caused widespread property and infrastructure damage. When Hurricane Delta arrived 5 weeks later it reexposed and reflooded buildings and homes in various states of disrepair. The additional water and wind damages created more complex insurance and federal assistance claims scenarios that event-based reporting systems were ill equipped to handle, resulting in delays in the flow of resources to municipalities and households alike, further disrupting livelihoods, social connections, health, and education. Less than 5 months later, Winter Storm Uri[7] froze the tarps on roofs, prompting another round of retarping; pipes froze and burst, but went unreported because many residents were still evacuated. Municipal water pressure remained low, and remaining

---

[7] This report uses the term "Winter Storm Uri," but the committee acknowledges that no official U.S. agency names winter storms. The name "Uri" is a construct of a private media company associated with the Weather Channel.

residents went without running water for almost a week. These stressors intensified the risk of disaster impacts for all residents, but exponentially intensified risk for populations with high exposure and vulnerability. Exhaustion, anger, and emotional numbness continued to compound the mental health of residents and was put to the test again when extreme rainfall flooded the area all over again 3 months later. Debris not yet removed from the 2020 storm sequence clogged stormwater systems, limiting the capacity available to absorb or drain the record rainfall. As a state official noted: "You don't have a break. It just keeps hammering." These increased complexities warrant an evolution of our understanding of disasters, particularly as it pertains to disaster recovery, which typically cannot be measured in months, but rather in years, decades, or even lifetimes.

Increased Risk of Compounding Disasters in the Gulf of Mexico Region

Climate change will continue to affect every region on Earth in a multitude of ways, with adverse impacts such as the severity, duration, and frequency of extreme weather-climate events increasing in the coming years (IPCC, 2021, 2022; USGCRP, 2023). Humans, animals, plants, and the ecosystems they inhabit will experience longer, hotter heat waves; longer summers and shorter winters; and intensification of the water cycle, accompanied by flooding and more intense rainfall in some regions and drought in others (USGCRP, 2023). Seas will continue to warm and rise, increasing the likelihood that hurricanes will continue to intensify, with damaging winds, and life-threatening storm surge. Climate change can also increase the probability of pandemics like COVID-19 and exacerbate their impacts (Ernst et al., 2023). Accordingly, climate change is an omnipresent, direct and indirect, threat to population health.

During 2020–2021, two phenomena—the COVID-19 pandemic and climate change—were shaping disaster risk worldwide, with particularly severe consequences in the GOM region. The emergence and spread of COVID-19 transformed the public health risk of all other extreme weather-climate events by amplifying health-compromising exposures and underlying vulnerabilities in the region. The pandemic also modified preparedness, response, and recovery procedures for the 2020–2021 weather-climate events. The combined result was a complex and unprecedented public health and socioeconomic crisis—an exemplar of compounding disasters.

Disasters are the instantiation of risk and are largely a social construct, influenced by human choices, priorities, and values (Oliver-Smith, 2022). Oftentimes, a natural hazard (e.g., a hurricane) is conflated with the impact the hazard can have when it intersects with vulnerable and exposed people and deemed to be a "natural disaster." Accordingly, the disasters discussed in this report are not referred to as natural disasters, despite the widespread use of that term. A hazard may be natural in origin and further amplified by human-caused climate change, but a disaster is characterized by the exposure and vulnerability of people and places rather than the hazard itself. The combination of climate-intensified hazards, a low-lying coast with major population centers, poor health outcomes, and degraded infrastructure result in high exposure and vulnerability within the GOM region, and are just several factors that make residents particularly susceptible or sensitive to disasters and even more so to compounding disasters (Burton, 2010; O'Keefe et al., 1976; Steinberg, 2000).

Compound(ing) Disaster Literature and Impacts in the Gulf of Mexico Region

The convergence of climate change and the global COVID-19 pandemic in 2020 further complicated conventional conceptualizations of disasters, spurring more publications relating to compound disaster events in the United States and globally. While disaster science has advanced in recent years, an understanding of the interrelatedness, connections, and complexities of exposures, vulnerabilities, and adaptive capacities that can determine the scale of compounding disasters diverges among disciplines and remains largely siloed (Kruczkiewicz et al., 2021).

A growing list of related terms—including *complex, simultaneous, concurrent, parallel, cascading, cumulative, multihazard, domino, conjoint, multivariate, preconditioned, connected, compound,* and *catastrophic*—continues to be used in the literature to describe disaster events without agreement on their definitions (Cutter, 2018). The literature on compounding risks, events, and disasters, inclusive of the multiple terms and conceptualizations, can provide a broader scientific context for the study and its findings and conclusions.

As sea level rise, drought, and other long-term stressors challenge conventional notions of a disaster, researchers are calling for more attention to gradual-onset effects (Tosun and Howlett, 2021; Yamori and Goltz, 2021) as well as to expand theoretical examinations of extreme event(s) attribution (Zscheischler and Lehner, 2022) and empirical evidence of compounding disasters' associated health threats, particularly for vulnerable populations (de Ruiter et al., 2020; Ebi et al., 2021; Gao and Wang, 2023; Singh et al., 2023). Repeated exposure to hurricanes can prompt an increase in mental health symptoms (Garfin et al., 2022), while anxiety and stress may develop in individuals experiencing cumulative fatalities/injuries from multiple disasters (Hu et al., 2021). Cascading disasters, such as Winter Storm Uri and the power grid failure, led to increased thoughts of suicide (Sugg et al., 2023).

Simpson and colleagues (2021) propose a risk assessment that incorporates a holistic view of climate change risks that recognizes mitigation and adaptation responses as potential risk drivers and that considers risk interactions (e.g., how multiple risks aggregate, compound, and cascade). Others are urging scholars and practitioners to integrate "multi-hazard thinking" or "compound thinking" into risk assessments, emergency management response plans, mitigation policies, and postdisaster settlement planning (Lee et al., 2024; Tsai et al., 2021; van den Hurk et al., 2023), as well as incorporating local and traditional knowledge (Nakashima et al., 2018).

The experiences of many GOM residents in 2020–2021 echoed many of the complexities described in the literature, confirming the evolution of the increasingly entangled nature of compounding disasters and the inadequacy of current disaster response and recovery approaches within the disaster management enterprise.

Complexities that were experienced ranged from incompatible response protocol (e.g., biological versus weather-related hazards; Potutan and Arakida, 2021); interactions between risks of transboundary climate extremes and slow-emerging climate change effects (e.g., sea level rise; Harris et al., 2022); attribution of direct/indirect disaster mortality counts that further complicated postdisaster mortality surveillance (NASEM, 2020; Santos-Lozada and Rivera-Reyes, 2024); and disaster experiences that produce compound disadvantages that accumulate across domains, resulting in a "multiplicity of impact" (Priest and Elliot, 2023).

Overlapping, Interrupted, and Prolonged Recoveries

In the GOM region, disaster recovery is a process that is unbound by time. Communities with high levels of social and economic vulnerabilities are more likely to experience prolonged or indefinite disaster recovery periods (e.g., disposal of improperly remediated disaster debris near low-income neighborhoods; Allen, 2007) and ongoing mental health impacts

(Cherry et al., 2021) during which time their sensitivity to experience the onset of each new disruptive event as a disaster is amplified.

In the GOM region, communities are increasingly living under "continuous threat, response, and recovery disaster cycles" (Osofsky et al., 2022, p. 2). In other words, while communities are responding to a disaster, they will also be recovering from a previous disaster (Lukasiewicz and O'Donnell, 2022), blurring the lines between the mitigation, preparedness, response phases, and extending recovery into what more closely resembles an epoch within a disaster survivor's lifetime. Within this undefined period of perpetual disaster recovery, risk and sensitivity to adverse impacts from a future disaster or disruptive event are heightened.

When multiple disasters overlapping in space and time occur, with residual effects persisting and the recovery time line extending indefinitely, it becomes more difficult to ascribe loss and damage to the occurrence of a particular discrete event. Yet, in seeking disaster recovery assistance from FEMA, applicants are required to submit separate applications for each disaster wherein applicants must "describe what damage happened on which date to ensure you're applying under the correct disaster" (FEMA, 2024).

Just as parsing the physical damage that occurred during each disaster can blur together, recoveries from these disasters also can become indistinct. A nongovernmental organization leader from Harris County explains how he experienced this reality in an information-gathering session: "In my line of work, rebuilding houses, we still haven't recovered from Harvey— potentially haven't recovered from the Tax Day and Memorial Day floods in '14 and '15. So all that recovery blends together."

The disaster recovery process is nonlinear, contingent on and bound by cultural histories and geographies (Hsu et al., 2015). The associated systems and policies both create and sustain barriers to recovery and the inequitable distribution of resources (Thomas et al., 2020), defined by universal or predetermined, inflexible parameters that can reproduce inequities and deepen vulnerabilities. Thus, while a disruptive event may be bound by a specific temporal time frame of onset and conclusion (e.g., the hurricane has passed or dissipated), the recovery period associated with the aftermath is far more challenging to encapsulate when interrupted by another disruptive event. It is through this lens that the committee sought to understand the factors that position communities, some more than others, to experience the compounding effects of multiple disruptive events and overlapping and interrupting disaster recoveries.

The Rise of Billion-Dollar Disasters

During 1980–2023, the costs of billion-dollar disasters in the United States topped $2.6 trillion (see Figure 1-3). Adding to this total the costs attributed to weather-climate events for which damages fell below the billion-dollar threshold could increase this amount by as much as 20 percent (NOAA, 2022b). Every U.S. state and territory is susceptible to costly weather-climate disasters, but three states in the GOM region—Florida, Louisiana, and Texas—currently rank highest in terms of disaster-related costs. Each state incurred more than $200 billion in damages between 1980 and 2023. Figure 1-3 shows total billion-dollar disaster costs by state during this period. Since 1980, Texas has led the nation in total cumulative costs resulting from billion-dollar weather-climate disasters ($402 billion). Florida currently ranks second ($389 billion) but was briefly supplanted by Louisiana ($304 billion) as a result of Hurricane Ida in 2021 (NCEI, n.d.-b; A. Smith, 2022).

Adjusting for inflation, billion-dollar disasters are steadily rising in number (NCEI, n.d.-a). In 1980, 3 disasters (all impacting at least one GOM state) met or exceeded the $1 billion threshold, compared with 28 in 2023 (NCEI, n.d.-a). Longitudinal examination of billion-dollar disasters over the decades reveals a steady increase in such events: 33, 57,

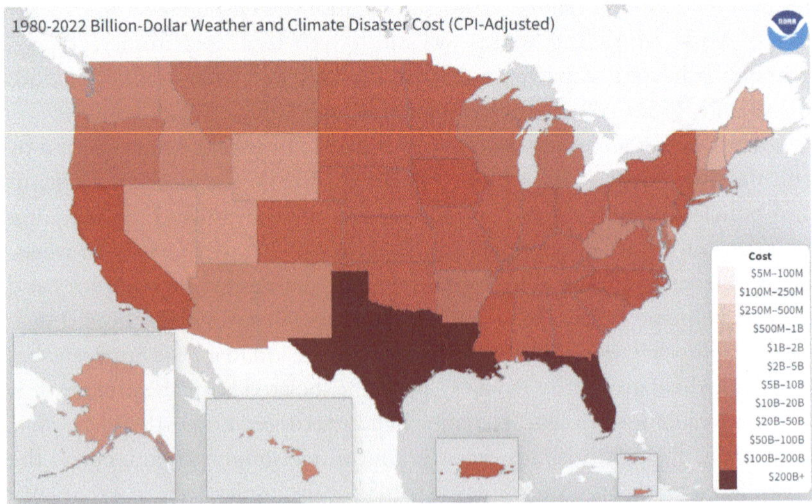

FIGURE 1-3 Total costs of billion-dollar disasters, 1980–2023, by state.
SOURCE: NOAA, 2024.

67, and 131 in the 1980s, 1990s, 2000s, and 2010s, respectively. In 2020, 22 billion-dollar (or greater) weather-climate disasters occurred in the United States (NCEI, n.d.-c); 17 of these took place in the GOM region (NCEI, n.d.-b). In 2021, 20 such disasters occurred (NCEI, n.d.-c), 15 in the GOM region (NCEI, n.d.-b), together resulting in approximately 950 direct or indirect fatalities (NOAA, 2022b).

From January 1, 1980, to September 11, 2023, GOM states experienced 243 billion-dollar disasters (see Figure 1-4), which equates to an annual average of 5.5 such events in that time frame. In the 5-year period from 2019 to 2023, this annual average more than doubles, to 14.6 events (NCEI, n.d.-a). Overall, these 243 disasters resulted in the official, direct deaths of 11,121 residents (NCEI, n.d.-a). During the 2-year span of 2020–2021 GOM states collectively endured 32 billion-dollar disasters, including six hurricanes in 2020 (Delta, Eta, Hanna, Laura, Sally, and Zeta) and in 2021, two tropical storms (Elsa and Fred), two hurricanes (Ida and Nicholas), a major flood event, a hailstorm, and a severe winter storm (Uri).

Multiple challenges exist when comparing recent and historical billion-dollar disasters. Inflation is one consideration, but increases in population and building activity must also be taken into account. For example, if there were a $700 million event in 2013, adjusting for inflation, that event would not reach the billion-dollar mark today. However, if a disruptive event of the same magnitude as the 2013 event occurred in an area with

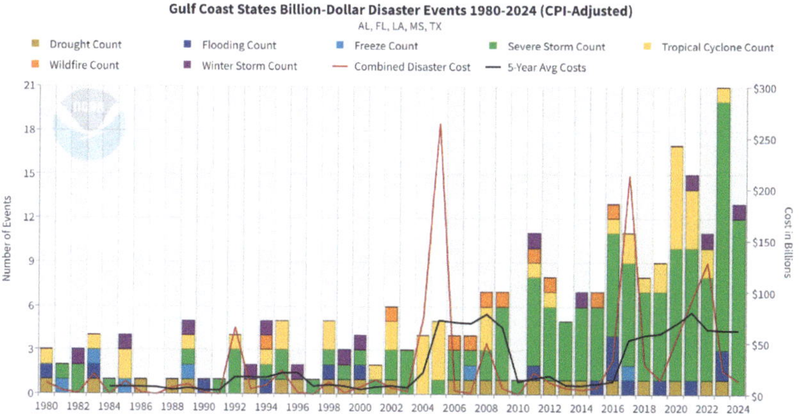

FIGURE 1-4 Type, number, and total cost of GOM states' billion-dollar disasters, 1980–2023.
SOURCE: NCEI, 2023.

increased population and building construction, it would likely be deemed a billion-dollar event because of the greater exposure of people and buildings (Theodorou, 2023).

Disasters Hidden Beneath the Billion-Dollar Disaster Threshold

The federal disaster declarations and billion-dollar designations for the major storms of 2020–2021 do not account for the less well-known or unnamed storms or disruptive events that occurred during those 2 years. Such events contribute to accumulated losses and damage to communities lacking infrastructure and proper mitigative measures—features that often align with vulnerability factors such as income, race, and rurality (Hendricks and Van Zandt, 2021).

Geographically limited and lower-magnitude events such as low-scale tornadoes, flooding, and even tropical systems experienced in the GOM region do not meet the threshold of a billion dollars, nor do they receive the attention as major disasters tagged with such superlatives as "deadliest," "largest," and "earliest." Yet, they are disruptive events that can stress communities and their undergirding systems and further prolong recovery periods from previous disruptive events and disasters. In parts of Lake Charles, Louisiana, for example, a heavy rain event in May 2021 produced 12.49 inches of rainfall in 24 hours, 6 inches of which fell in a 2-hour period (Di Liberto, 2021), while many residents were still reeling from 2020's Hurricanes Laura and Delta and Winter Storm Uri. This was the third-highest amount of rain at one time on record in Lake Charles (Benedict, 2022), and it caused significant flooding, creating obstacles for a region already suffering from compounding disasters. The sensitivities of the affected population elevated the disruptive event to compounding disaster status.

The baseline for GOM states—one of extreme risk driven by vulnerability, upon which growing hazard exposure is overlaid—exacerbates the potential for future compounding disasters. Mitigating the effects of such disasters will require addressing the drivers of vulnerability, exposure and uneven migration, while concurrently increasing understanding of the multiple, interacting risks and responses to climate change (Simpson et al., 2021) and compounding disasters.

RELEVANT NATIONAL ACADEMIES REPORTS

The National Academies has produced numerous reports focused on strategic planning related to disaster preparedness and on health and resilience related to climate change. A number of these reports are relevant to the work of this committee, and this report builds on the findings, conclusions, and recommendations of those earlier studies. Highlights of these reports are presented in Box 1-2.

BOX 1-2
Relevant National Academies Reports

In *Climate and Social Stress: Implications for Security Analysis* (NRC, 2013), the authoring committee contends that, as the frequency and severity of environmental hazards increase with climate change, the nation will likely begin to see "clusters of apparently unrelated climate events occurring closely in time" and "events in which a climate event precipitates a series of other physical or biological consequences" (p. 3). Such climate events can precipitate social disruptions in both the affected local area and beyond, in some cases causing shocks to global supply chains. The committee concludes that a robust assessment of the likelihood of intersecting threats, coupled with "timely preventive measures" and effective emergency response, will be necessary if the United States is to prevent and respond to potentially catastrophic events (p. 139).

In *Disaster Resilience: A National Imperative* (NRC, 2012), the authoring committee acknowledges that "disasters will continue to occur . . . in all parts of the country" and that "impacts of climate change and degradation of natural defenses such as coastal wetlands make the nation more vulnerable" (p. 14). The committee recommends that federal agencies incorporate resilience as a guiding principle at all levels and that both the public and private sectors "work cooperatively to encourage commitment to and investment in a risk management strategy that includes complementary structural and nonstructural risk-reduction and risk-spreading measures or tools" (p. 212).

In *Healthy, Resilient, and Sustainable Communities after Disasters: Strategies, Opportunities, and Planning for Recovery* (NASEM, 2015), the authoring committee identifies the importance

continued

BOX 1-2 Continued

of human health and equity in rebuilding resilient communities after a disaster, stating that "equity, resilience, and sustainability are all relevant to post-disaster efforts to build healthier communities" (p. 52). The committee specifically recommends that health considerations be integrated into disaster recovery plans at the local, state, and federal levels. Additionally, the committee identifies access to "affordable, high-quality, and location-efficient housing" as an essential element of any healthy community (p. 51). The committee specifically discusses the necessity of (re)building public and accessible housing suited to low-income, elderly, and disabled community members in order to create equitable disaster recovery.

In *Attribution of Extreme Weather Events in the Context of Climate Change* (NASEM, 2016), the authoring committee attempts to elucidate the extent to which an "individual event's magnitude or probability of occurrence" can be attributed to anthropogenic climate change (p. 2). According to the committee, while it is usually not possible to conclude definitively whether anthropogenic climate change caused an individual weather event, evidence suggests that climate change has influenced the frequency and severity of contemporary heatwaves, heavy rainfall, consecutive dry days, droughts, tropical cyclones, and other extreme events. For more information on attribution science, see the "Hazards" section of Chapter 2 of that report.

In *Building and Measuring Resilience: Actions for Communities and the Gulf Research Program* (NASEM, 2019a), the committee catalogs and evaluates the various efforts to measure community resilience in the Gulf of Mexico (GOM) region and offers suggestions for communities to build their community resilience according to the metrics identified. The committee first identifies six types of community capitals that can provide a holistic view of resilience: natural (or environmental), built (or infrastructure), financial (or economic), human and cultural, social, and political (institutional or governance; pp. 15–16). The committee suggests that communities should take four "key actions" to build their resilience: "1. Build community engagement and buy-in, 2. Account for communities' multiple dimensions, 3. Link community resilience measures to decision making, and 4. Create incentives for measuring resilience" (p. 60). The committee concludes that community decision-makers must work alongside researchers in longitudinal research and resilience planning to develop resilience initiatives that will result in long-term success.

In *A Framework for Assessing Mortality and Morbidity After Large Scale Disasters* (NASEM, 2020), the authoring committee describes the current challenges with accurately assessing and reporting mortality and morbidity following large-scale disasters and offers recommendations to improve mortality and morbidity data collection and reporting after disasters. According to the committee, two components of the problem with postdisaster mortality and morbidity reporting are that (a) there are multiple and complex health outcomes that can result from disasters and (b) there is no one measure to quantify the impact of every disaster, and a singular death toll cannot capture the full scope of disaster health impacts. The committee recommends that the U.S. Department of Health and Human Services not only count direct disaster fatalities, but should also work with communities to calculate individual count and population-level estimates of postdisaster mortality and morbidity that distinguish between direct deaths or morbidities, indirect deaths or morbidities, partially attributable deaths and morbidities, and unrelated deaths and morbidities, which are defined in the report (p. 170).

In "Opinion: Compound Risks and Complex Emergencies Require New Approaches to Preparedness," Kruczkiewicz and colleagues (2021) describe how the COVID-19 pandemic has further complicated response to acute disaster events. The authors identify the problem thusly:

> The ability to anticipate and respond is constrained by a lack of available resources at the right place at the right time, limited governance and accountability, and an underestimation of uncertainty in forecasts for both climate and societal impacts, including social, economic, political, and infrastructural. The COVID-19 crisis further decreases disaster resilience and thus increases baseline risk and the potential scale of impacts on systems, lives, and livelihoods, which in turn increases vulnerability to future disasters (p. 3).

The authors argue that "flexible funding mechanisms and strategies" distributed among "diverse cooperative networks" of stakeholders will be essential to allow communities to plan for and recover from multiple concurrent hazards and disaster events effectively (p. 4).

In *Resilience for Compounding and Cascading Events* (NASEM, 2022a), the authoring committee describes a "new normal" in which disasters occur in rapid succession, "often unleashing new devastation on a community before it has had a chance to recover from the prior disaster" (p. 3). The committee acknowledges that climate change makes the occurrence of compounding disasters increasingly

continued

BOX 1-2 Continued

likely. The committee calls for more extensive research on compounding and cascading disasters to help "develop solutions and avoid unintended consequences" that might occur within the current model of emergency management (p. 25).

In *Engaging Socially Vulnerable Communities and Communicating About Climate Change–Related Risks and Hazards* (NASEM, 2022b), the authoring committee writes that socially vulnerable populations are especially vulnerable to the impacts of extreme weather-climate events:

> It is important to recognize that information and communication are asymmetrical, and vulnerable groups—including pregnant people, children, older adults, immigrant groups, Indigenous peoples, low-income communities, communities of color, people with disabilities, vulnerable occupational groups (e.g., workers who are exposed to extreme weather), and people with preexisting or chronic medical conditions, among others—often suffer from a comparative lack of authoritative information (American Public Health Association [APHA], 2022; USCRP, 2018).
>
> The extreme weather events exacerbated by climate change also often have disproportionate impacts on the above vulnerable populations (APHA, 2022; Dietz, Shwom, and Whitley, 2020; IPCC, 2022; Jacobs, 2019; Méndez, Flores-Haro, and Zucker, 2020; USGCRP, 2018). Among these impacts are psychological distress; physical impairments; and loss of income, property, and livelihood (Wong-Parodi, 2020). Moreover, extreme weather events exacerbated by climate change can compound existing vulnerabilities and inequities experienced by these populations, stemming from past and present structural and systemic discriminatory practices that can limit the ability of communities to adapt (Chen et al., 2022; Jacobs, 2019; USGCRP, 2018) (pp. 2–3).

The committee suggests that, to communicate more effectively with socially vulnerable populations, decision-makers should "form partnerships with trusted and diverse community organizations" (p. 4), develop a structured decision-making process between community members and policymakers, address the social vulnerabilities that cause such inequities, and "allow and encourage community ownership and leadership of responses" (p. 6).

In *Advancing Health and Resilience in the Gulf of Mexico Region: Roadmap for Progress* (NASEM, 2023a), the authoring committee identifies the GOM region as especially socially and ecologically vulnerable and describes the need for more effective ways to meet community needs:

> The Gulf of Mexico region, in particular, suffers from recurrent disasters that have reduced its physical infrastructure, separated communities from resources, and exacerbated the challenges required to deliver the right services, in the right place at the right time, and in the ways that are most beneficial to the communities. The development of a holistic, systems approach to meeting the health and community resilience needs of people living in the Gulf region will be dependent on a planful and deliberate effort that aligns the service delivery needs with both appropriate investments in physical infrastructure and the human and material resources necessary to carry it out, with appropriate accountability and sustainability built in (pp. 100–101).

The committee provides several recommendations directed at research funders and state and federal partners on how to develop more holistic methods for advancing health and resilience in GOM communities, including designing resilience initiatives around community needs with community participation; engaging local community members in research opportunities; and promoting "health in all domains," such as "mental health care, education, housing, and social infrastructure" (p. 107).

Finally, in *Community-Driven Relocation: Recommendations for the U.S. Gulf Coast Region and Beyond* (NASEM, 2024), the authoring committee chronicles the history and contends with the present reality of communities affected by acute and chronic impacts of climate change, particularly sea level rise and flooding. The committee examines the experience of socioeconomically vulnerable and marginalized communities and offers recommendations for addressing more equitably the priorities and needs of those who have been or may be displaced from their homes in pursuit of greater physical safety.

ORGANIZATION OF THE REPORT

This report is intended for an audience of decision-makers at the institutional and community levels interested in expanding their understanding of the impacts of compounding disasters on communities; how and why effects compound and amplify; and what factors enabled or could enable communities to successfully plan for, respond to, recover from, and mitigate the impacts of compounding disasters.

Chapter 2 describes the disaster risk profile of the GOM region and lays out GOM hazards and baseline exposure and vulnerability conditions, including factors that amplified the impacts of compounding disasters in 2020–2021. Chapter 3 provides an account of the 2020–2021 compounding disasters and summarizes some of what the committee heard in the public information-gathering sessions in Harris and Galveston Counties, Texas; Calcasieu and Cameron Parishes, Louisiana; and Mobile and Baldwin Counties, Alabama, as it relates to participants' experiences of compounding disasters. Chapter 4 uses a systems approach to analyze how the effects of compounding disasters interacted with the interconnected and interdependent systems critical to societal functioning, as well as connective factors that may have amplified those impacts. The committee selected this approach because it offers an integrative framework with which to disentangle underlying and emerging patterns, relationships, causality, and feedback loops involved in the 2020–2021 disasters (e.g., the COVID-19 pandemic's co-occurrence with other disruptive events). Chapter 5 summarizes and examines lessons recognized, learned, and implemented, during the 2020–2021 time frame, in addition to lessons learned and lost from prior disasters. Throughout the narrative content of each chapter, significant committee findings are identified in bold typeface. A summary of findings is also provided at the conclusion of each chapter. Finally, Chapter 6 presents the committee's conclusions as to the factors that enabled or could enable communities to successfully plan for, respond to, recover from, and mitigate the impacts of compounding disasters.

2

Hazards, Exposure, Vulnerabilities, and Disaster Risk in the Gulf of Mexico Region

Driven predominantly by use of fossil fuels and associated carbon emissions, climate change is a global existential threat to communities. Climate change drives slow- and rapid-onset changes in the physical environment, prompting disruptive and deleterious social changes with lasting health, community, and economic impacts. In short, **climate change is the overarching threat multiplier for risks to population health, safety, and well-being in the Gulf of Mexico (GOM) region**. Moreover, social structures and networks, including systemic behaviors, prejudice, racism, and institutionalized policies, result in inequitable distribution of climate change impacts, resources, and delivery of services (Singh et al., 2023).

A range of social, economic, geographic, demographic, cultural, environmental, historical, political, and governance factors (i.e., social determinants of health; see Box 2-1) influence exposure and vulnerability to hazards, including those associated with climate change, and poor health outcomes. Relatedly, inequities in education, income, access to transportation, affordable and safe housing and/or neighborhoods, and health care and social services drive disproportionate vulnerability and exposure to disruptive events. These inequitable circumstances, particularly among communities of color, Indigenous populations, and those living in poverty and areas of historical socioeconomic disadvantage, increase disaster risk and constrain opportunities to expand adaptive capacity, community resilience, and health and prosperity, further elevating the risk of future disaster impacts.

BOX 2-1
Social Determinants of Health in the Gulf of Mexico Region

Nested within the variables of vulnerability and exposure are multiple social determinants of health, defined as the conditions in the environments where people are born, live, learn, work, play, worship, and age that affect a wide range of health, functioning, and quality-of-life outcomes and risks (ODPHP, 2023). The U.S. Department of Health and Human Services groups social determinants of health into five domains: economic stability, education access and quality, health care access and quality, neighborhood and built environment, and social and community context (ODPHP, n.d.), which represent dimensions of both vulnerability and exposure in the context of disaster risk.

Baseline conditions for social determinants of health are generally poorer for Gulf of Mexico states than for the rest of the United States, with systematic disparities in mortality and other measures of well-being even across small areas that lie relatively close together (NASEM, 2017). These conditions worsen during disaster response and recovery, heightening inequities and sensitivity to subsequent disruptive events (Martin, 2015). The response and recovery phases of a disruptive event also elevate the risk of such outcomes as poverty, illness, mental health struggles, housing instability, social isolation, death, and exposure to threats or violence, limiting opportunities to recover effectively (Drabek, 2007; Smiley et al., 2018). Adverse social determinants of health thus can render communities unevenly and profoundly exposed and vulnerable to adverse impacts and heightened challenges at every disaster stage (Bikomeye et al., 2021).

Population exposure to disasters has been documented as producing a range of consequences for physical and mental health, as well as the environment. Disasters impact public health by causing damage, destruction, disruption of health care services, displacement, debility, disease, and death. They can exacerbate preexisting mental health conditions for individuals, but on a broader population level, contribute to stress and distress that may lead in some cases to disaster event–related, new-onset mental disorders. When compounding disasters occur, public health impacts may be amplified by simultaneous or sequential exposure to harmful impacts along with the experience of loss and disruption of life in the aftermath of the events (Hahn et al., 2022; Leppold et al., 2022; Wells et al., 2022).

Against this backdrop, this chapter explores the drivers of disaster risk in the GOM region: hazards, exposure, and vulnerabilities beginning with a brief background on climate attribution science and a description of the natural and technological hazards that contributed to compounding impacts in the GOM in 2020–2021. Next, it provides an overview of exposure associated with land-use patterns, past and current, in the GOM region, that have placed populations, infrastructure, housing, production capacities, and other tangible human assets in hazard-prone areas. Finally, it describes extant physical and social vulnerabilities in the GOM region wherein disaster impacts are "piled on" and disaster risk is exacerbated. The broad impacts of disasters on public health are discussed as well as the disproportionate risks social vulnerability has created for population health in the GOM region. Impacts to health system performance and the psychological footprint of a disaster follow, with a discussion of medically vulnerable patient populations. Social capital and cohesion is also discussed, followed by GOM public infrastructure and building inventories. Next is a discussion on housing's role in vulnerability, including the U.S. legacy of racial discrimination in homeownership opportunities; displacement following disasters; and disproportionate impacts on renters, rental homes, and manufactured housing units. Disasters' impacts on the changing insurance market is followed by a discussion of economic instability and inequities before and during the period of a presidentially declared disaster.

HAZARDS

Hazards in the GOM region include extreme weather-climate events, land subsidence, and technological hazards related to fossil fuels, all of which were compounded during 2020–2021 by the COVID-19 pandemic.

Climatologically, the GOM region is susceptible to a variety of hazards, including winter storms, wildfires, drought, and extreme temperatures; however, hurricanes (or tropical cyclones), flooding, and severe local storms (including tornadoes) are the most common weather threats. Hurricanes produce significant impacts on coastal and inland communities, including such hazards as storm surge, heavy rainfall, inland flooding, and strong winds (Ashley and Ashley, 2008; Fussell et al., 2017; Muller and Stone, 2001; Rappaport, 2000). Storm surge occurs when hurricane winds push seawater onto shore, causing flooding, erosion, and property damage. Heavy rainfall can cause inland flooding, while strong winds can damage buildings, power lines, and other infrastructure (Henry et al., 2020). In addition, hurricanes can result in significant economic losses, including damage to crops, businesses, and transportation systems (Faber, 2015; Schmidt et al., 2009). Hurricanes and other hazards are occurring within a changing climate amid a dynamic demographic and economic landscape, driving numerous complex and interrelated challenges, ongoing community disruptions, and enduring adverse impacts.

Attribution Science

Attribution science attempts to disentangle the influence of human-caused climate change in individual extreme weather-climate events from other factors, such as the climate patterns El Niño and La Niña (NASEM, 2016). Attribution studies (Fowler et al., 2021; NASEM, 2016; van der Wiel et al., 2016) confirm that the influence of climate change is represented by contemporary heatwaves, heavy rainfall, consecutive dry days, droughts, tropical cyclones, and other weather-climate events. However, because the science is relatively new,[1] it is still an area of active research (NASEM, 2016), and attribution results are more robust for certain events (e.g., heatwaves) than for others (e.g., increased extreme precipitation). **Recent studies suggest that the occurrence of rapidly intensifying and stronger tropical cyclones may be increasing as a result of climate change** (Knutson et al., 2021; Walsh et al., 2016). In the Atlantic Basin, the number of major hurricanes (Category 3, 4, or 5) has increased, and there is some evidence that hurricanes are intensifying, slowing down, or stalling as they approach the coast (Hall and Kossin, 2019; Kossin, 2018; Walsh et al., 2016).

Shepherd and colleagues (2007) document that rainstorms associated with lower-category hurricanes are the most prolific rain producers, with inland freshwater flooding being one of the deadliest aspects of landfalling tropical cyclones (Rappaport, 2014). Hurricane Harvey (in 2017) is a good example of how a "stalled" storm can produce disastrous flooding, winds, and other hazards. In Harris County, Texas, Smiley and colleagues (2022) showed that 30 to 50 percent of Harvey-flooded properties would not have flooded without climate change using climate attribution science paired

---

[1] The first publication attempting to attribute an extreme weather-climate event to climate change was in 2004, analyzing the 2003 European summer heat wave (see Stott et al., 2004).

with hydrological models, and detailed land-parcel and census tract socio-economic data.

Witze (2018) demonstrates how climate warming is leading to "wetter" storms, which likely include tropical cyclones (and their remnants). The Clausius-Clapeyron relationship clearly establishes that a warmer atmosphere has more water vapor capacity (Martinkova and Kysely, 2020). Balaguru and colleagues (2018) and Bhatia and colleagues (2022) also found evidence of increasing trends in rapid intensification of tropical cyclones, a phenomenon that hampers the ability to prepare and evacuate adequately for landfalling storms. Some studies also suggest that climate change could lead to more polar vortex weakening events (Cohen et al., 2021), such as those that spurred Winter Storm Uri. Reed and colleagues (2022) demonstrated that human-induced climate change played a role in the full 2020 hurricane season by increasing storm rainfall rates and accumulated rainfall amounts for observed storms that were at least tropical storm strength.

While climate is shaped by far more complex processes (Balaguru et al., 2023; IPCC, 2021; USGCRP, 2023), a simplification of the effects of climate change as related to wind, warming air, and water is offered here. As humans burn more fossil fuels, ocean and atmospheric temperatures grow warmer. The warmer the ocean temperature, the higher the volume of water and the more evaporation is available to the atmosphere. The warmer the atmosphere, the more moisture it can hold, and the more rain can be produced. More rain means more released heat, and more heat means stronger winds. Wind speed is nonlinearly related to potential structural damage. The damage potential from a Category 4 storm with 150 mph winds (the strength of Hurricane Laura at landfall) is 256 times greater than that from a Category 1 storm with 75 mph winds. Stronger storms also mean the potential for higher storm surge, due not only to stronger winds driving storm surge inland, but also to rising ocean temperatures that lead to an expansion of the sea's volume and corresponding sea level rise (Barlow and Camargo, 2022).

Despite the relative nascency of attribution science, valuable data can be gleaned from this research to inform future risk calculations pertaining to extreme weather events and their impacts on critical infrastructure, housing, food management, building codes, and insurance (NASEM, 2016). Studies have shown that these data can even be used to establish liability for adverse climate-related impacts for litigation purposes (Burger et al., 2021; Burton, 2010; CRS, 2023).

Hurricanes and Tropical Storms (Tropical Cyclones)

Since official records began in 1851, 216 hurricanes have made landfall between the Rio Grande River in Texas and the Florida Keys, 80 of which were rated as major hurricanes (NOAA, n.d.-a). Hurricanes and tropical storms have wide-ranging regional impacts, affecting both coastal regions and areas further inland. They are often rated on the well-known Saffir-Simpson wind scale, but an array of associated hazards is possible, including storm surge inundation, strong winds, tornadoes, intense rainfall, and pluvial (surface water) and fluvial (waterbody overflow) flooding (Andersen and Shepherd, 2013; Fussell et al., 2017; Rappaport, 2000). These hazards threaten life, safety, physical and mental health, private and public property, household incomes, residential stability, public infrastructure, and utilities.

The GOM region houses vital energy, industrial, and transportation infrastructure. Wind damage, storm surge, and flooding compromise or disable power transmission, energy production, and road networks (Henry et al., 2020; Schmidt et al., 2009). Figure 2-1 shows the number of tropical cyclones per decade in each county and parish in the Southeast United States.

Since 1980, the time between landfalling tropical storms in the GOM region has shortened (Xi and Lin, 2021) and this pattern may be escalating as a result of climate change (Knutson et al., 2010; Walsh et al., 2016). This phenomenon is particularly evident in the Atlantic basin, where the number of major hurricanes has increased in recent years (Walsh et al., 2016). Hurricane hazards at landfall (e.g., strong winds, heavy rainfall) are also expected to continue (Xi et al., 2023). Additionally, the dynamics of hurricanes are changing. Their rapid intensification reduces the time available for preparations and evacuations (Bhatia et al., 2022; Wang et al., 2019). Evidence also suggests that some hurricanes are slowing down, which results in increased rainfall and flooding (Kossin, 2018; Patricola and Wehner, 2018) and higher storm surge (Marsooli et al., 2019). **These shifting climate conditions will likely contribute to continued changes in hazard characteristics: in combination with the heavily populated and built-up coast, an increase in disruptive events, subsequent compounding of disasters, and likely less time for disaster recovery before the next tropical cyclone makes landfall.**

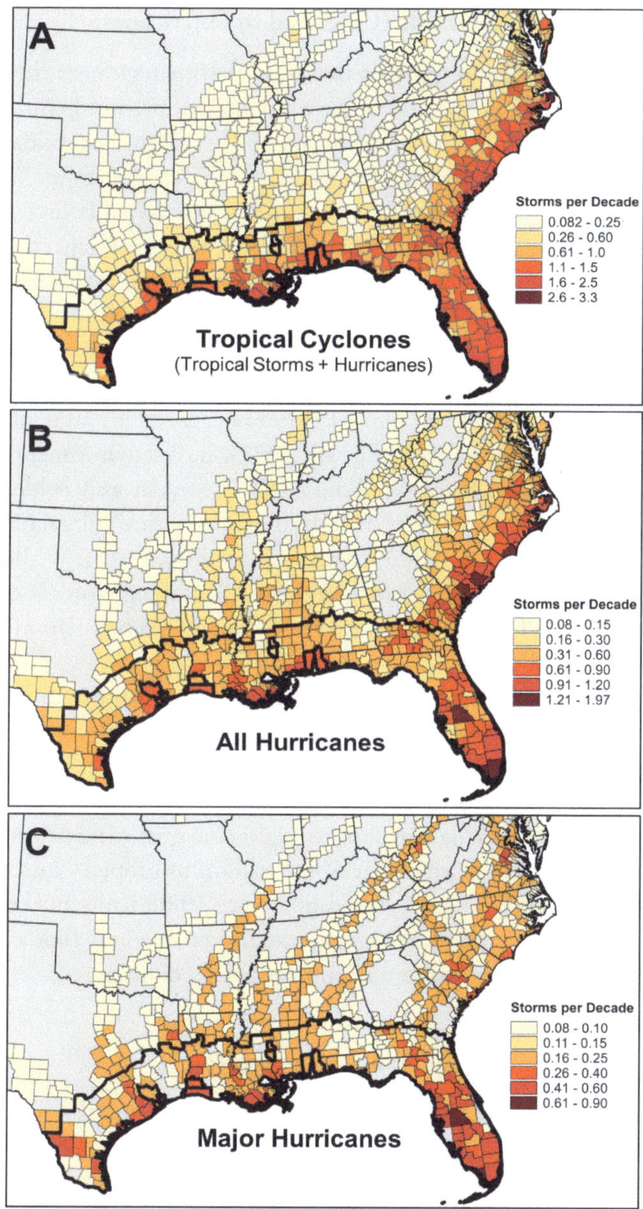

FIGURE 2-1 Frequency of county/parish-level landfalling storms per decade since 1900: (A) tropical storms and hurricanes, (B) all hurricanes (Category 1–5), and (C) major hurricanes (Category 3–5). Saffir-Simpson scale ratings were determined by the storm intensity at landfall.
SOURCE: Strader, 2023.

Wind Hazards (Unrelated to Hurricanes)

The GOM states are part of the Southeast, a region of the United States exposed to tornadoes and other wind hazards predominantly during periods of active thunderstorms in late spring, with a secondary peak in late fall. While the Central Plains are often referred to as "Tornado Alley," recent studies have found that tornado activity is increasing in the Southeast (Gensini and Brooks, 2018), which experiences a greater frequency of significant (EF2+ [Enhanced Fujita Scale]) tornadoes, as well as a larger overall tornado damage footprint (Ashley and Strader, 2016; Dixon et al., 2011).

Tornado-related fatalities are also more likely in the Southeast than in the rest of the continental United States because of an array of factors, including storm motion (Strader et al., 2022), built-environment density, the nocturnal nature of storms (Ashley, 2007; Strader and Ashley, 2018; Strader et al., 2022), and greater social and infrastructure vulnerability (Ash et al., 2020; Cutter et al., 2003; Strader and Ashley, 2018). While having a smaller damage footprint compared with landfalling tropical cyclones, tornadic storms kill a greater percentage of the exposed population (Ashley, 2007; Strader et al., 2022).

Tornadic activity is not the only source of wind damage in the GOM region. Murley and colleagues (2021) found that high-wind events, or HWEs, meeting National Weather Service criteria are most prevalent in the South and West North Central regions. The authors note that such events, which may include derechos, straight-line gust winds, downbursts, or microbursts, have caused nearly $300 million in property losses during the past 60 years and roughly 1,500 deaths since 1980. Knox and colleagues (2011) have documented the role of non-convective winds (not associated with a hurricane or convective storm) in property damage.

Extreme Temperatures

Since global temperature records began in 1880, Earth's temperature has risen by an average of 0.14°F (0.08°C) per decade, or about 2°F in total, with the 10 warmest years occurring since 2010 (Lindsey and Dahlman, 2023). From 1981 to the present, the warming rate per decade has been more than double the rate of 0.32°F (0.18°C) as previous decades (Lindsey and Dahlman, 2023), and 2023 recently surpassed 2016 as the hottest year on record (Copernicus, 2024).

Marine heat waves (when the ocean temperature is warmer than 90 percent of the previous observations for any given time of year) are occurring in GOM waters, stressing sensitive marine ecosystems (NOAA, 2023), raising heat and humidity on nearby land during the hottest times of the year and providing "jet fuel" for tropical cyclones (Harvey, 2024).

GOM states currently lead the nation in numbers of days of perilous heat. Climate change, humidity, low elevation, and warm waters in the GOM will contribute to an average of 20 extra days of triple-digit heat per year (First Street Foundation, 2022). Residents of Texas and Florida can expect to see more than 70 consecutive days with the heat index surpassing 100°F (37.7°C) (Muyskens et al., 2023). Three GOM states have the second (Texas), third (Mississippi), and fourth (Louisiana) highest rates of extreme heat.

Extreme temperatures and weather-climate events are among the leading causes of major power outages in the United States, prompting increased electricity demand for heating and/or cooling, which can stress and overload aging energy infrastructure (Climate Central, 2024). From 2018 to 2020, three GOM states had the highest (Louisiana), second-highest (Texas), and fourth-highest (Mississippi) annual average counts of 8+ hour outage events. For 1+ hour outage counts, these states rank highest (Texas), second highest (Louisiana), and third highest (Mississippi) (Do et al., 2023).

Hazards associated with landfalling tropical cyclones are often exacerbated and compounded by power outages. Roughly 90 percent of major power outages in the United States are associated with hurricanes (Alemazkoor et al., 2020). Since tropical cyclones are inherently warm season hazards, the impact of heat in the aftermath of a storm is often understudied. In fact, very few studies are found in the literature. Reesman (2022), in a thesis focused on Hurricane Laura and defoliation, found evidence of increased apparent temperature during nocturnal hours after the storm. Guido et al. (2022) reported that heat index anomalies surged in the days following many hurricanes in the Caribbean region.

Sea Level Rise and Coastal Flooding

By 2050, sea level along U.S. coastlines is expected to rise by 10–12 inches (0.25–0.30 m; see Figure 2-2 for relative sea level rise estimates for 2050 and 2100 under the Intermediate Sea Level Rise Scenario), resulting in extensive erosion, land loss, and more frequent overland flooding in coastal communities (Dangendorf et al., 2023; NOAA, 2022c; Yin,

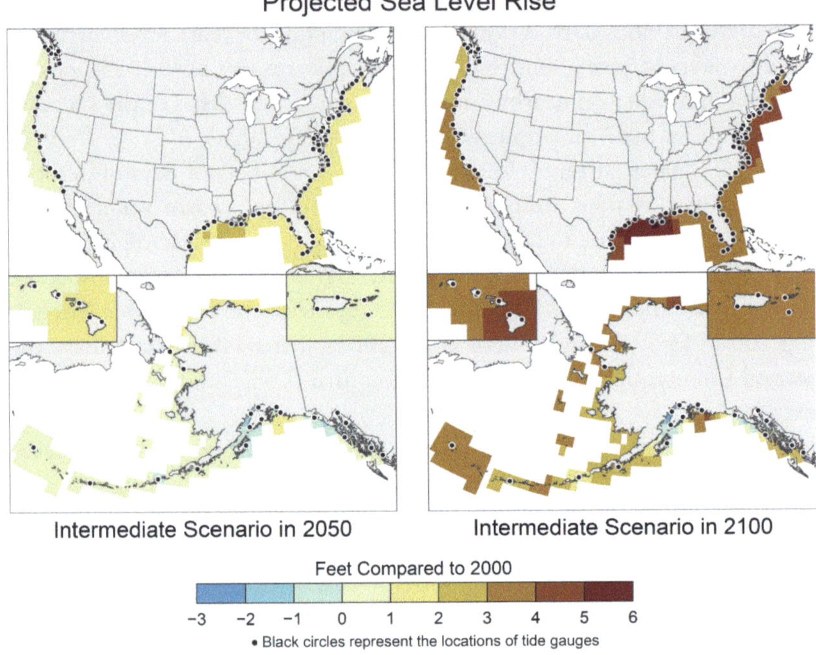

FIGURE 2-2 Relative sea level rise along the U.S. coastlines under the Intermediate Sea Level Rise Scenario of the U.S. Interagency Sea Level Rise Task Force for 2050 (left) and 2100 (right). By both 2050 and 2100, sea level rise is projected to be highest along the western Gulf Coast.
SOURCE: USGCRP, 2023.

2023). Notably, sea level rise rate projections in the United States are highest along the western Gulf Coast (see Figure 2-2).

Land Subsidence

Subsidence, the gradual sinking of the land surface, results from several natural and anthropogenic factors. The Mississippi River Delta Plain naturally subsides as sediments compact under the weight of overlying material. Levees have greatly reduced the rejuvenation of sediment to the delta surface, and consequently subsidence exceeds land building. Other factors include oil, natural gas, and water extraction, which contribute to the sinking of the land mass. Groundwater removal has been a primary factor in Texas. Faulting in the delta sediments also plays a role (Lane et al., 2018; USGS, n.d.). Recent research reveals that subsidence rates vary across the Gulf Coast, but some areas are experiencing subsidence at a rate of 4 mm per year (Wang et al., 2020; Zhou et al., 2021). Although marshes can sustain themselves during moderate sea level rise, Törnqvist and colleagues (2020) report that the Mississippi Delta marshes are nearing a tipping point for long-term sustainability. In coastal marshes, land subsidence amplifies the effects of sea level rise. Sea level rise and land subsidence are phenomena that, like other chronic, slow-moving hazards such as "nuisance flooding" and the slow march of drought, can have significant agricultural, environmental, economic, health, and social consequences on their own, but in combination with more acute stressors and a changing climate, can be catastrophic.

Technological Hazards: Chemical Releases

The GOM region is the primary U.S. hub for the fossil fuel industry. Forty-eight percent of total U.S. petroleum refining capacity is situated in the GOM region (USEIA, 2023a) with more than 90 percent of U.S. primary petrochemicals capacity located in Texas and Louisiana (Augustine, 2024). Together, Louisiana and Texas have the 10 largest petrochemical complexes in North America and produce more than a third of the United States' methane (USEIA, 2023a). Many residents, disproportionately communities of color and with low socioeconomic status live close to these and other high-risk chemical facilities, including transport facilities and/or carcinogen-laden hazardous waste (Superfund) sites. Many of these communities have been overburdened by the environmental harms and risks from exposure, cumulative impacts, disproportionate health impacts, and greater vulnerability to pollution for decades (Saha et al., 2024; Terrell and James, 2022; see "Exposure to Environmental Contaminants" later in this chapter for more information).

There are more than 4,000 oil and gas structures (NOAA, n.d.-b) in the GOM region, interconnected by more than 26,000 miles of oil and gas pipelines (NOAA, n.d.-c), making the region vulnerable to contaminant spills, leaks, and explosions (e.g., 2010 *Deepwater Horizon* disaster). Human health hazards are evident at every point along the petroleum production pathway, from extraction (drilling or hydraulic fracturing ["fracking"]) to refining to petrochemical production to burning to end-product production (e.g., plastics) to disposal and management of wastes. Liquefied natural gas (LNG) facilities, for example, emit several hazardous air pollutants

(HAPs) that are known to cause a variety of human health problems (Saha et al., 2024). Some of these HAPs include "formaldehyde (a known carcinogen that causes myeloid leukemia and nasal cancers), benzene (also a known carcinogen), toluene (which can cause a range of reproductive harm including birth defects), ethylbenzene (a suspected carcinogen that can cause hearing and kidney damage), and xylene (which has wide-ranging effects)" (Saha et al., 2024, p. 172). According to the Federal Energy Regulatory Commission, six LNG facilities located in Texas and Louisiana are anticipated to collectively emit 264 tons of these HAPs annually (Saha et al., 2024).

Chemical releases occur on blue-sky days as well as in the context of a disruptive event. Natural Hazards Triggering Technological Accidents (Natech) events are incidences of natural hazards (e.g., flooding) that initiate events that challenge the safety and operation at hazardous installations (OECD, 2024). Technological failures and Natech events are not only increasing in the United States (Sengul et al., 2012) but are expected to continue to rise in both frequency and magnitude (Krausmann et al., 2017) as climate conditions worsen (Saha et al., 2024). Yet, they are typically overlooked in regional and national disaster risk management plans (Girgin et al., 2019).

The COVID-19 Pandemic

With extraordinary speed, COVID-19 circumnavigated the globe, moving from the Western Pacific to worldwide over the span of 1 month—March 2020 (Shultz, Perlin et al., 2020). The World Health Organization declared COVID-19 to be a "public health emergency of international concern" in January 2020 (WHO, 2020a) and then upgraded it to "pandemic" status in March 2020 (WHO, 2020b). In the United States, the pandemic was declared a national emergency under Section 501(b) of the Stafford Act (Robert T. Stafford Disaster Relief and Emergency Assistance Act, 42 U.S.C. 5121 et seq.) on March 13, 2020. The sudden emergence and ongoing evolution of the COVID-19 pandemic joined climate change as phenomena of such global scope and scale as to fundamentally influence concurrent disaster events at all levels: worldwide, nationally throughout the United States, and with disproportionate impact in the GOM region due to the confluence of the COVID-19 variants during the 2020 and 2021 hurricane seasons. As a hazard, the COVID-19 pandemic stressed the public health system, including those who work and/or volunteer in the sector at the local, state, and regional levels. The pandemic has also heightened the health care sector's awareness that social conditions influence health (Gottlieb et al., 2021).

EXPOSURE

Exposure to extreme weather-climate hazards in the GOM region is increasing as a result of population growth and increased construction of public and private development in hazard-prone areas. In addition to potential increases in the temporal frequency or rate of return of extreme weather-climate hazard events, several studies have found that migration and development near the U.S. Gulf Coast are also contributing to escalating losses from these events (Hoffman, et al., 2023; Pielke and Landsea, 1998). *Exposure* is understood as land development that places people, infrastructure, housing, production capacities, and other tangible human assets in hazard-prone areas. The term encompasses the biophysical factors that may contribute to a disruptive event (e.g., residence in a coastal area or a river floodplain). Figure 2-3 illustrates how the expansion of the built environment can increase exposure to flood hazards and thereby increase risk associated with a flood disaster. Growing exposure along the Gulf and Atlantic Coasts is linked to mounting hurricane losses (Freeman and Ashley, 2017; Zhu and Quiring, 2022). Much of this construction is also inherently *vulnerable* because of inadequate construction standards, discussed later in this chapter. (See also Strader, 2023, for more information on how the GOM societal landscape is shaping cyclone and tornado disasters.)

Population and Population Exposure

Of the three U.S. census-designated coastline regions (Pacific, Atlantic, and GOM),[2] the GOM is the smallest coastline region by area but the fastest growing in terms of overall population (U.S. Census Bureau, 2019). From 2000 to 2017, the GOM coastal population increased by 26.1 percent; by context, the U.S. population growth rate over the same time frame was 15.7 percent (U.S. Census Bureau, 2019).

From 1940 to 2020, GOM states experienced significant population growth (432.7 percent, 34 million people) and corresponding exponential growth in housing units (2,818 percent, 17 million homes; Strader, 2023). Of the six GOM representative counties/parishes that are a focus of this

---

[2] Counties or parishes adjacent to coastal water or territorial sea.

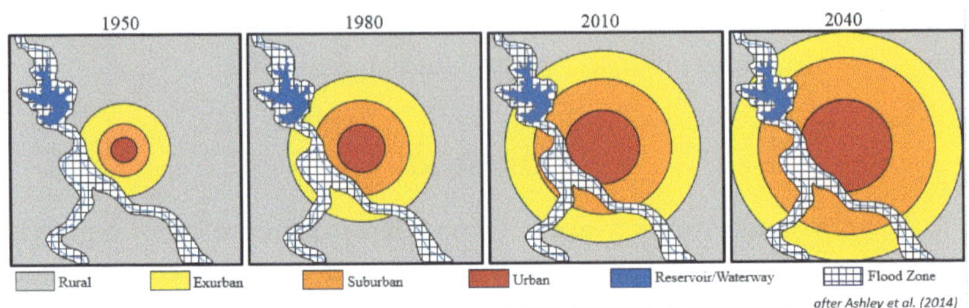

FIGURE 2-3 The "expanding bull's-eye effect" is a conceptual model for a hypothetical metropolitan region characterized by increasing development spreading from an urban core over time. A sample flood scenario is overlaid to illustrate that flood risk increases as population and built environment expand into areas in or near flood zones.
SOURCE: Ashley et al., 2014.

study, Harris County, Texas, experienced the highest rate of growth in population (32.3 percent) and housing units (61.1 percent) from 1940 to 2020, driven largely by the Houston metro area. Baldwin County, Alabama, and Calcasieu Parish, Louisiana, also saw significant increases in population (20–30 percent) during that period, with corresponding increases in housing units (approximately 50 percent). Of note, vacation or secondary residences comprise many of the housing units in the coastal study area. Vacation homes account for 39.7 percent of the housing units in Baldwin County, Alabama, 63.2 percent of the housing units in Cameron Parish, Louisiana, and 40 percent of the housing units in Galveston County, Texas, as of 2022 (U.S. Census Bureau, 2020).

Rapid population growth in GOM states has further contributed to extending and expanding persistent socioeconomic inequities (USGCRP, 2023), pushing many low-income residents toward low-lying areas with flood risks amplified by climate change and increased runoff from expansion of the built environment (Ueland and Warf, 2006).

The 2020–2021 period was also marked by sharp population shifts. From April 2020 to May 2022, Texas and Florida saw among the highest U.S. population increases in terms of relative growth (first and fourth, respectively, in percent growth). Regionally, within the GOM states of Louisiana and Mississippi, for example, there are notable exceptions. Population estimates released in 2021 for Cameron and Calcasieu Parishes in Louisiana show that these localities experienced some of the nation's sharpest declines and the largest combined percentage decrease compared with metro areas nationwide (M. Smith, 2022). This decline is part of a wider trend in Louisiana, as well as in Mississippi, which ranked fifth and ninth, respectively, in relative decline (third and seventh, respectively, in percent decline) (U.S. Census Bureau, 2022c). It must be noted that, as populations depart the coast and parish/county budgets decline, inadequate funding will become an increasing concern. Selected census-derived 2020 demographics for the six representative counties/parishes that are a focus of this study and Marion and Jefferson Davis Counties, Mississippi, are shown in Table 2-1.

Exposure to Wind Hazards

A population's exposure to wind hazards may best be characterized by the hazard maps in ASCE 7, an American Society of Civil Engineers standard that underpins the International Codes.[3] ASCE 7-22 establishes hurricane-prone regions as those areas with basic wind speeds of 115 mph or higher, which are subject to heightened design requirements (ASCE, 2022a). This region extends from the coastline of the GOM states to approximately 100–150 miles inland. Within this zone, ASCE 7-22's wind speed contours signify increasing exposure to strong winds by progressively raising the basic wind speed a building must resist the closer it is to the coastline. For example, structures in ASCE 7-22 Risk Category II, which would include noncritical, low-occupancy buildings such as single-family homes, built in Galveston, Texas, are expected to experience wind speeds of 150 mph, just slightly higher than the basic wind speeds in Mobile, Alabama (146 mph) and markedly higher than the 129 mph basic wind speed requirements in the inland city of Lake Charles, Louisiana (see Figure 2-4).

These standards presume that hurricanes weaken as they pass over land. Yet fast-moving storms like Hurricane Laura challenged this assumption and delivered winds in excess of the basic wind speed assumed in ASCE 7 well inland of the coast, including in Lake Charles (Roueche et al., 2020). ASCE 7-22 also defines tornado-prone regions of the United States, which includes the entirety of all GOM states, highlighting that geographies within 100–150 miles of the GOM coastline face heightened exposure to all varieties of wind hazards compared with the overall continental United

---

[3] The International Codes are developed by the International Code Council, which uses a governmental consensus process to develop International Codes (I-Codes) (ICC, 2024). I-Codes are the minimum standard for building design to ensure safety well-being (ICC, 2024).

TABLE 2-1 Selected Demographics of Counties/Parishes in GOM States in 2020

County, State	Population	Race–White alone (%)	Race–Black or African American alone (%)	Ethnicity–Hispanic or Latino (%)	Owner-occupied housing unit rate (%)	Median income ($)	Poverty level (%)
Baldwin, AL	231,767	83.4	8.4	5.0	77	61,756	9.2
Mobile, AL	414,809	55.7	36.7	3.2	64.1	44,091	17.6
Cameron, LA	5,617	88.1	3.9	4.6	88.5	56,902	13.1
Calcasieu, LA	216,718	67.2	24.8	4.4	68.5	54,530	17.4
Marion, MS	29,341	65	31.5	1.8	75.9	40,978	17.3
Jefferson Davis, MS	11,321	37.9	59.1	1.9	80.3	32,214	22.7
Harris, TX	4,731,145	69.6	20.1	43.7	54.9	63,022	15.6
Galveston, TX	350,682	80.3	13.3	26.8	67.5	74,633	10.9
U.S.	331,464,948	58.9	13.6	19.1	64.8	75,149	11.4

NOTE: Searches conducted for galvestoncountytexas, cameronparishlouisiana, calcasieuparishlouisiana, marioncountymississippi, jeffersondaviscountymississippi/FIPS.
SOURCE: U.S. Census Bureau, 2021a.

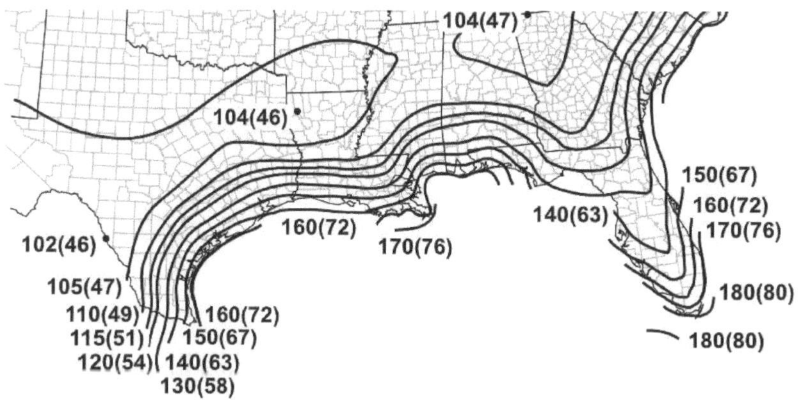

FIGURE 2-4 Basic wind speeds for Risk Category II buildings and other structures along the Gulf Coast.
SOURCE: ASCE, 2022a, with permission from ASCE.

States. Thus, exposure to the damaging effects of wind can persist for miles inland from the coast. This wind hazard exposure includes not only the direct impacts of the wind pressure on the built environment but also heightened exposure to other related hazards, such as windborne debris and wind-driven rain, both of which contribute significantly to losses often outside of delineated flood zones.

Exposure to Coastal Hazards

More than 18,130 km² (7,000 mi²) of the Gulf Coast is 1.2 m (4 ft) or less above current sea levels. This area includes critical infrastructure (e.g., ports, airports, hospitals, schools, petrochemical facilities, interstates, railroads) essential to local and national economies and large urban centers (e.g., Houston and New Orleans; Strader, 2023). Approximately 25 percent of health care facilities and public safety assets and 20 percent of schools in all six counties/parishes that are a focus of this study are flood prone (within 100 m of a 100-year floodplain; Strader, 2023). In Galveston County, Texas, more than 60 percent of hospitals are flood prone, as is the only hospital in Cameron Parish, Louisiana, where 99.7 percent of surface roadways are in a flood hazard zone (Strader, 2023). Exposure to hazards is heightened by the fact that substantial portions of the housing inventory are flood prone. Of the six counties/parishes, Cameron Parish (93.3 percent) and Galveston

County (49.2 percent) contain the highest percentages of their building footprints in flood-prone areas (Strader, 2023).

Sea level rise and flood exposure will significantly impact businesses and residents adjacent to or dependent upon the bayous, coastlines, and coastal waterways, and also inland areas subject to riparian flooding (Colten, 2021), as people migrate away from the rising water and subsiding land. Figure 2-5 illustrates physical factors that can shape the potential for exposure to flood risk.

Exposure to Extreme Temperatures

Extreme cold temperatures are decreasing in frequency and intensity globally (Hu et al., 2020), while extreme heat temperatures are increasing (Song et al., 2022). But both heat and cold have measurable health effects, particularly among vulnerable populations (Adams-Fuller, 2023; Jay et al., 2021; USGCRP, 2023), and are the deadliest weather-climate-related phenomenon in the United States (Adams-Fuller, 2023; NOAA, 2021n).

In the United States, extreme heat kills approximately 1,300 people every year—more than hurricanes, tornadoes, and floods combined (Adams-Fuller, 2023). Thirty-year projections for the number of days with a heat index of 100°F (37.78°C) or higher show a worsening situation in the coming decades (Adams-Fuller, 2023). Within the next 30 years, the dangers of extreme heat are projected to be the most widespread in the southern United States. As climate change continues to amplify exposure risk to extreme heat and hot weather, the associated health stress is escalating mortality, morbidity, adverse pregnancy outcomes, and adversely affecting mental health and constraining the ability to work outdoors (Ebi et al., 2021).

Despite the known risk, this hazard receives minimal dedicated federal support and funding for planning, education, mitigation, and recovery (Wickerson, 2023). Decision-makers and policymakers and the public tend to focus on more episodic or telegenic events (e.g., landfalling hurricanes, tornadoes), in part because extreme heat events are ineligible for major disaster declaration under the Stafford Act, precluding deployment of resources and coordinated action, and as such, the consequences of extreme heat are difficult to respond to. Extreme temperatures (thus far) largely do not damage property or cause as much physical destruction as severe storms yet are due far more attention in the face of global temperature warming trends.

With planning, education, and action, heat-related impacts are preventable (NIHHIS, n.d.). Localized and well-communicated heat action

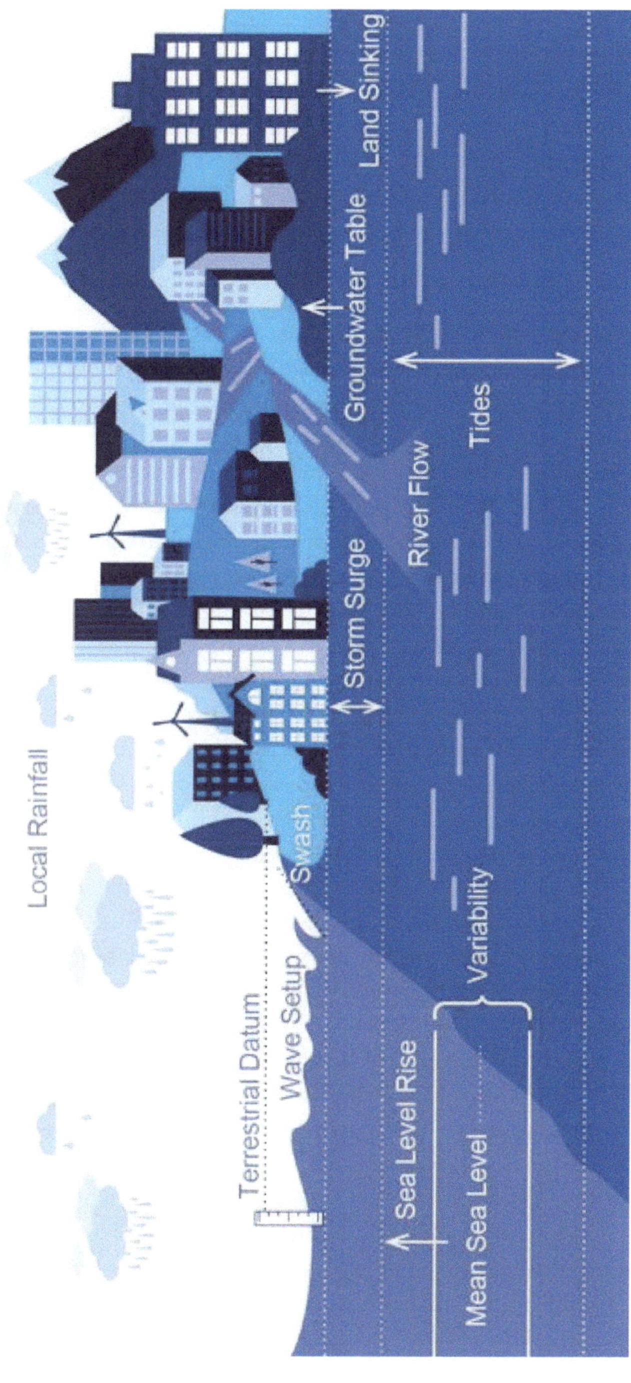

FIGURE 2-5 Physical factors contributing directly to flood exposure.
SOURCE: NOAA, 2022e.

plans can help elevate public discourse on the ways in which extreme heat is and will continue to interrupt daily activities. Plans that address behavioral approaches and biophysical adaptations as well as include surveillance, monitoring, and evaluation can reduce the interruptions and adverse health consequences of current and future extreme heat (Ebi et al., 2021; Errett et al., 2023; Jay et al., 2021).

Exposure to Environmental Contaminants

One of the most consistent findings in the social science literature is that *place matters*. More than two decades of research show a relationship between the location of environmental hazards (e.g., petrochemical production and transport, by-products associated with hydraulic fracturing, landfills, hazardous waste sites) and racial (e.g., African American) and/or economic (e.g., low-income) status. **GOM populations are disproportionately and unequally exposed to environmental contaminants, toxicants, and other hazards that not only exacerbate negative physical and mental health outcomes on blue-sky days** (Domingue, 2022; Donaghy et al., 2023; Prochaska et al., 2014), **but also pose increased risks to human health when hurricanes, flooding, and other weather-climate events mobilize these toxicants to enter communities.**

Furthermore, because many industrial activities often require access to surface water, industrial land often overlaps with flood zones (Flores, Castor et al., 2021), increasing the susceptibility of adjacent neighborhoods to hazards associated with the petrochemical industry (Bernier et al., 2017). This increased exposure occurred, for example, during and after Hurricane Harvey, when many low-income communities were disproportionately exposed to chemical contaminants (Karaye, Stone et al., 2019) wherein a U.S. Environmental Protection Agency Office of Inspector General report (U.S. EPA, 2019) concluded that neither the state of Texas nor local authorities adequately monitored air quality during or immediately following the storm. Nor did these entities effectively communicate with residents about the risks posed by the chemical releases and explosions. Further, many industrial facilities were caught off guard and were forced to shut down and restart, resulting in the release of an estimated 2,750–4,150 tons of excess emissions (Phillips, 2018). Flooding prompted more chemical releases, including an explosion at the Arkema chemical plant in a suburb of Houston that burned for 4 days after a backup generator failed (CSB, 2017).

Increased exposure also regularly occurs in Louisiana, which ranks among the bottom five states for air pollution and mortality (Yu et al., 2023). In a region commonly known as "Cancer Alley," where more than half (378 of 671) of the state's industrial facilities are spatially clustered along a 184-mile stretch of the lower Mississippi River, communities of color have "7-fold to 21-fold higher emissions, depending on the pollutant, than predominantly White communities" (Terrell and St. Julien, 2023). Chemical manufacturing facilities are disproportionately responsible for pollution emissions in Cancer Alley, accounting for about one-third of emissions of particulate matter (PM10 and PM2.5), nitrogen oxides (NO_x), sulfur dioxide (SO_2), carbon monoxide (CO), and volatile organic compounds (Terrell and St. Julien, 2023). Exposure to these chemicals can contribute to a wide range of negative health impacts such as chronic obstructive pulmonary disease (COPD), asthma, bronchiolitis, lung cancer, cardiovascular events, central nervous system dysfunction, and cutaneous diseases (Manisalidis et al., 2020). Residents of some Cancer Alley towns were forced to abandon the tradition of porch sitting in the evening due to nighttime chemical releases (Baurick, 2019).

Despite the higher levels of exposure to environmental risks and health inequities, these communities have fewer resources to address the adverse effects of environmental contaminants (Shepherd and KC, 2015; U.S. EPA, 2021; USGCRP, 2023). Understanding of these inequities resulted in the concept of environmental justice and fenceline communities—those located alongside environmental hazards—which experience disproportionately higher rates of cancer (Raun et al., 2013), asthma (Byrwa-Hill et al., 2023), cardiovascular diseases (Motairek et al., 2023), and COVID-19 deaths relative to the general U.S. population (Fos et al., 2021; Terrell and James, 2022). Fenceline communities' proximity to industrial facilities means that they are not only exposed to potentially toxic emissions but face higher risk of exposure to a Natech event (Nicole, 2021; see "Technological Hazards: Chemical Releases" earlier in this chapter for more information on Natech events).

VULNERABILITY

Vulnerability is shaped by complex forces, and is a result of the range of institutional, political, economic, social, cultural, and psychological factors and systems that configure people's lives and the environments in which they live (UNDRR, n.d.-a). Disparate yet interdependent social stratification processes (Kuran et al., 2020) influence the degree or sensitivity to which an individual or community is vulnerable to and affected by disruptive events. *Sensitivity* encompasses the social factors that lessen the ability of a person, family, or community to cope with a disruptive event, and can determine the degree to which they are affected by such an event.

***Adaptive capacity* can mitigate the potential for harm by reducing sensitivity to a disruptive event.** Adaptive capacity enables a human community to adjust to environmental conditions and effectively offset vulnerabilities (Smit and Wandel, 2006).

The identities, conditions, and demographics of vulnerable individuals can be intersectional, operating together to increase disaster risk. As Tierney (2019, p. 127) states, people "are not born vulnerable, they are made vulnerable." Vulnerable groups include, but are not limited to, pregnant people (Sharma et al., 2022), women (Nguyen et al., 2023), ethnic/racial minorities, children, elderly adults, Indigenous people, low-income communities (Bullard and Wright, 2012), undocumented immigrants, unhoused populations, prisoners, people with disabilities (Shultz et al., 2018), LGBTQIA+[4] populations (Haworth et al., 2022), unprotected occupational groups (e.g., workers exposed to extreme weather; Kiefer et al., 2016), medically high-risk patients (Balbus and Malina, 2009; Espinel et al., 2022; Shultz et al., 2024), and the politically marginalized (Joseph et al., 2021). Vulnerabilities and the composite burden of multiple, interrelated sensitivities are place based, and the resulting consequences are contingent on the particular disaster or disasters. Actions to address them must be tailored to the complexities of particular situations that arise from local populations and social, economic, and political conditions.

Many people who are vulnerable along one dimension (e.g., experiencing housing instability) are vulnerable along others as well (e.g., having difficulty accessing health care or employment opportunities). These individuals experience intersecting, overlapping, clustered, and at times compounding vulnerabilities even before a disruptive event occurs (Stanley, 2017). The impacts of the event are then "piled on" to these extant vulnerabilities.

Some factors that contribute to vulnerability are based on non-modifiable risk factors (e.g., age is unalterable, yet children and elderly adults bear disproportionate disaster burdens). This same concept applies

---

[4] LGBTQIA+ is an abbreviation for lesbian, gay, bisexual, transgender, queer, intersex, asexual, and other identities (Haworth et al., 2022). These terms are used to describe a person's sexual orientation or gender identity.

to discrimination based on gender, race, and/or ethnicity; women, for example, are 14 times more likely than men to die in disasters (CDP, 2023; Howe, 2019; Neumayer and Plümper, 2007). Persistent social vulnerabilities, disparities, inequalities, and inequities are dominant social forces that shape population health, local economies, the education system, and infrastructure quality, and ultimately exacerbate disaster risk.

Many vulnerabilities relate directly to sensitivity to hazards. For example, the density of mobile/manufactured housing can lead to greater odds of dying in a tornado, particularly in the Southeast United States; young and elderly populations are more likely to be killed; and those living below the poverty line have less ability to withstand impacts and losses (Strader, 2023).

High Social Vulnerability in Gulf of Mexico States

Communities throughout the GOM region lead the nation across a spectrum of vulnerability measures. Since 2000, social vulnerability has increased in the region, especially for the percentage of persons or households that are unemployed (+3.7 percent), aged 65 or older (+3.6 percent), and/or members of minority groups (+5.5 percent; Strader, 2023).

Several GOM states rank among the highest in the nation on, among others, the Centers for Disease Control and Prevention's (CDC's) Social Vulnerability Index (SVI).[5] Table 2-2 presents county-level SVI vulnerability measures for GOM states in 2000 and 2018, as well as the percentage change for each variable for the GOM region and the United States.

Social vulnerability in the GOM region continues to be approximately 25 percent higher than that in the rest of the continental United States (Strader, 2023). GOM states rank high on poverty measures and low on household income and educational attainment (see Box 2-2 for information on the role of education in disaster risk and recovery). This substrate of concentrated disadvantage has substance, depth, and constancy that reinforces a negative feedback loop and elevates risk for the subset of GOM communities as they grapple with disruptive events.

Because social vulnerability is socially constructed (Trivedi, 2023), context specific, and an outcome of demographic and historical legacies intrinsic to a place (Bankoff, 2004; Kuhlicke et al., 2011), some have expressed caution as to how single composite indexes, such as the SVI,

---

[5] The CDC's SVI is a place-based index, database, and mapping application designed to identify and quantify communities experiencing social vulnerability (CDC, 2024).

TABLE 2-2 County-Level Vulnerability Measures from the Social Vulnerability Index (SVI) for the GOM States and Percent Change, 2000–2018, GOM States and United States

Theme	Variable	Year 2000 (%) (GOM)	Year 2000 (%) (U.S.)	Year 2018 (%) (GOM)	Year 2018 (%) (U.S.)	2000 to 2018 Change (%) (GOM)	2000 to 2018 Change (%) (U.S.)
Socioeconomic Status	Below Poverty	19.0	14.2	19.9	15.6	0.9	1.1
	Unemployed	3.7	3.4	7.4	5.8	3.7	2.1
	Per Capita Income ($)	16,027	17,513	24,045	27,036	8,018	9,515
	No High School Diploma	29.0	22.6	18.3	13.4	−10.7	−9.2
Household Composition + Disability	Aged 65 Years or Older	14.0	14.8	17.6	18.4	3.6	3.6
	Aged 17 Years or Younger	26.0	25.5	22.8	22.4	−3.2	−3.1
	Civilian w/ Disability	24.0	20.9	17.0	15.9	−7.0	−5.1
	Single-Parent Households	10.1	8.2	9.9	8.3	−0.2	0.1
Minority Status + Language	Minority	36.0	18.7	41.5	23.5	5.5	4.8
	Speak English "less than well"	2.6	1.6	2.7	1.7	0.1	0.1
Housing Type + Transportation	Multi-Unit Structures	4.0	4.1	4.4	4.7	0.4	0.6
	Mobile/Manufactured Homes	24.0	14.9	22.0	12.9	−2.0	−2.1
	Crowding	5.9	3.6	3.3	2.4	−2.6	−1.2
	No Vehicle	9.1	7.6	6.8	6.4	−2.3	−1.2
	Group Quarters	4.3	3.4	4.6	3.5	0.3	0.1

NOTES: County-level SVI data for the years 2000 and 2018 show how individual vulnerability measures and the four primary SVI categories have changed over the 18-year period. Although 2020 SVI data are available, the categories used were slightly different from those used in 2000. Data from 2018, before the categories changed, are therefore used for the comparison with 2000. Pink cells indicate increasing vulnerability, and blue cells indicate decreasing vulnerability.
SOURCE: Strader, 2023.

BOX 2-2
The Role of Access to and Quality of Education in Disaster Risk and Recovery

Studies show that individuals with higher education experience healthier lives and longer lifespans (Raghupathi and Raghupathi, 2020) and are better positioned to build generational wealth (Kent and Ricketts, 2021). However, access to and quality of education take on increased importance in the context of disaster risk and recovery. In one longitudinal study (Thiede and Brown, 2013) on race, socioeconomic status, and evacuation behavior during Hurricane Katrina (in 2005), researchers found that people with less than a high school education were less likely to evacuate. Comparison of low-education respondents with respondents having at least some college education showed that the low-education respondents were more than twice as likely to have been unable to evacuate because of a lack of money, transportation, a place to go, or job requirements.

In another study (Tracy et al., 2011), conducted several months after Hurricane Ike, the researchers interviewed 658 adults who had been living in affected areas during the storm. The study showed that depression was more likely to occur among those with lower annual household income and fewer years of education (a high school degree or equivalent versus than those with some college or more years of education) (Tracy et al., 2011).

Nationwide, almost 90 percent of students complete high school, and many go on to higher education; however, all Gulf of Mexico (GOM) states have graduation rates well below the national average. After California (the state with the lowest high school graduation rate in the nation), three GOM states, that is, Texas, Mississippi, and Louisiana, have the second-, third-, and fourth-lowest rates, respectively, while Alabama has the seventh-lowest rate. Here too, there are disparities between racial and ethnic minorities and the non-Hispanic White population. While graduation rates continue to improve among racial/ethnic groups, the historical injustices and lack of access to quality education persistently place these groups at a disadvantage with respect to improving and increasing generational wealth. Limited education, low literacy, and a lack of broadband access can also hinder people's ability to understand and access disaster and other types of federal assistance.

are used (Roy and Berke, 2022). Marino and Faas (2020, p. 33) write, "There is a growing discomfort that categorizing the 'vulnerable' acts to flatten and simplify diverse communities, as well as discursively nullify the everywhere-visible 'resilience,' toughness, and genius that exists in communities, and subsets of communities, that are habitually exposed to risk." Quantitative indexes are only one metric for building an understanding of the social impacts of compounding disasters, and pairing indexes with empirical validation is one way to make them more robust (Rufat et al., 2019). Gathering information on people's lived experiences goes beyond the demographics of the people and places at risk and can elucidate the ways in which social marginalization—the root force that drives most forms of vulnerability—can prolong the impacts of compounding disasters (Priest and Elliott, 2023).

Many GOM residents are marginalized in terms of income; education; age; ethnicity/race; representation in policy, governance, and recovery planning; gender; sexuality; and/or being medically high-risk. These individuals and groups bear a disproportionate risk for and burden from compounding disasters (Bullard and Wright, 2012; Lichtveld, 2018; Machlis et al., 2022). **Their marginalization reduces their individual and collective adaptive capacity and magnifies their sensitivity to the consequences of future compounding disasters.** Comprehensive community participatory planning for future disruptive events is one way to help address these vulnerabilities before an event occurs. Participatory planning processes seek to engage and empower community members, especially marginalized people and those disproportionately affected by disasters to proactively engage in planning processes that build close partnerships with local and regional governments and strengthen community leadership and local adaptive capacities. Participatory planning can build and reinforce trust and enhance transparency and inclusivity via multiple levels of information gathering and sharing, while also strengthening social cohesion. As a component of participatory planning, there are tools and frameworks that communities can use to evaluate their adaptive capacities (see, e.g., Masterson et al., 2014; Notre Dame Global Adaptation Initiative; Sherrieb et al., 2010) This evaluation is one way to understand a community's strengths, assets, resources, and deficits to better inform community-based planning. Providing flexible resources—including training and financing—directly to community-based organizations and communities will help them assess their adaptive capacity, prepare and plan for, absorb, recover, and ultimately reduce impacts from ongoing and future disruptive events.

The recent Climate Vulnerability Index highlights where flexible resources and action are urgently needed for communities. It offers climate risk and vulnerabilities data at the census tract and county/parish level and incorporates input from community leaders. It comprises 184 indicators grouped under four baseline vulnerabilities (health, social/economic, infrastructure, and environment) and three climate change risks (health, social/economic, extreme events) to explain cumulative effects on neighborhood-level stability. Seven of the 10 most at-risk U.S. counties/parishes are in GOM states; of those, half are in Louisiana (EDF, 2023). Three GOM states—Texas, Louisiana, Alabama—and one southern state—Tennessee—account for 87 of the highest 100 scores of at-risk census tracts (Lewis et al., 2023). Of those, many are located near concentrations of industrial facilities and are known environmental justice communities (Lewis et al., 2023).

Material losses experienced by vulnerable communities in hazard-susceptible areas (e.g., health care, education, community infrastructure, housing, transportation) trigger and magnify existing social and spatial health stratifications (Weden et al., 2021), creating a vicious cycle that many cannot escape, and leaving some communities unable to prepare for the next disruptive event because of insufficient recovery from previous ones (Ingham et al., 2022).

Public Health

Disasters have broad and profound impacts on public health (Nomura et al., 2016; Shoaf and Rottman, 2000; Shultz, 2019; Shultz, Espinel, Galea et al., 2007), producing harms to individuals and communities in such forms as injury, disease, psychological distress and trauma, and death (Shultz, Espinel, Galea et al., 2007; Shultz et al., 2013, 2017; UNDRR, n.d.-b). The complexities and/or accumulation of multiple concurrent or sequential disaster exposures produce risks to population health that exceed those associated with single disaster exposures (de Ruiter et al., 2020; Ebi et al., 2021; Leppold et al., 2022; Lowe et al., 2020; Pescaroli and Alexander, 2018; Zscheischler et al., 2018). Disasters also may damage health care facilities and disrupt access to health care services, placing heavy demands on frontline health professionals (Shultz and Forbes, 2014)[6]

---

[6] Frontline health professionals include first responders, public health professionals, and volunteers tasked with medical and public health–related disaster response and recovery responsibilities.

and social support services (Heagele and Pacquiao, 2019). Each disaster has unique impacts on public health related to its distinguishing features (Shultz, 2019; Shultz and Neria, 2013; Shultz et al., 2015) and the underlying vulnerabilities at every social level—individual, family, social network, community, city, state, nation, and beyond.

Many residents of GOM communities suffer from poor health and health disparities that increase their vulnerability to disasters. Among key vital statistics, compared with those in the rest of the United States, life expectancies in the GOM region are low, while infant mortality and maternal mortality rates are high (CDC, 2023a, 2023b). Box 2-3 describes a study examining the relationship between compounding disasters and adverse health outcomes. The region also has some of the highest rates of incidence, prevalence, and mortality for the major non-communicable diseases that represent the leading causes of death in the United States, including heart disease, stroke, cancer, COPD, diabetes mellitus, and Alzheimer's disease (CDC, 2022; CMS, 2018). The clustering of very high mortality rates for these diseases in the GOM states of Alabama, Louisiana, Mississippi, and at times Texas is evident and generally consistent (CDC, 2022). The region also has a high prevalence of prominent lifestyle risk behaviors associated with these diseases, including physical inactivity, obesity, and cigarette smoking.

BOX 2-3
Compounding Disasters and Adverse Health Outcomes

A study by Hahn and colleagues (2022) examines the relationship between recent and compounding disasters and the incidence of acute and chronic health outcomes among residents in 500 U.S. cities at the census tract level from 2001 to 2015. Communities that had recently (within 5 years) experienced a disaster reported higher incidences of high blood pressure and asthma and worse mental health than communities that did not recently experience a disaster. The incidence of these poor health outcomes increased by 1–2 percent for each additional year a community experienced a disaster.

Population Health Disparities and Social Vulnerability

Research teams have documented how social vulnerability has created disproportionate risks for population health in the GOM region. Smiley (2020) examined racial inequalities in "flood extent," primarily outside the 100-year floodplains, during the unprecedented inundation associated with Hurricane Harvey. Using attribution science, Smiley and colleagues (2022) were able to demonstrate that 30 to 50 percent of flooded properties would not have flooded without climate change, and flooded properties were concentrated in low-income Latina/x/o neighborhoods primarily located outside of the Federal Emergency Management Agency's (FEMA's) 100-year floodplain.

Responding to the call to prioritize the integration of research into public health emergency response (Lurie et al., 2013), Horney and colleagues (2018) investigated residential contamination with polyaromatic hydrocarbons (PAHs) in household dust and outdoor soil in a Houston environmental justice neighborhood prior to and after Hurricane Harvey. Corroborating Horney et al.'s finding of redistribution of PAHs in low-socioeconomic status (SES) neighborhoods in the aftermath of Hurricane Harvey's flooding, Lieberman-Cribbin et al. (2021) found sharp socio-economic disparities when examining 83 toxic waste site releases during Harvey, with most toxic releases concentrated in neighborhoods in the lowest SES index quintiles. Sansom et al. (2023) and Atoba et al. (2023) documented poorer health indicators in residents living close to toxic release facilities and areas prone to flooding contamination.

Collaborators have also conducted in-depth research on the intersection of social vulnerability, health status, and disproportionate disaster impacts, with multiple studies focusing on Hurricane Harvey. Chakraborty, Collins, and Grinesky (2019) and Flores, Collins et al. (2021) bring an environmental justice lens to these analyses, illuminating that, prior to Harvey, racial/ethnic minority and low-SES populations were less prepared for disaster impacts and had limited resources to mitigate hurricane and flood hazards. This is important because better-prepared residents who were able to engage in pre-event mitigation had fewer physical health problems, post-traumatic stress symptoms, and adverse experiences (Grineski et al., 2020). This team conducted detailed analyses of disparate flood, toxic waste, and hazardous contaminant exposures during Harvey, linking these findings to pervasive social vulnerability (Chakraborty et al., 2021; T. W. Collins et al., 2019; Flores, Castor et al., 2021).

During Winter Storm Uri, Grineski, Collins et al. (2023) documented social disparities in the duration of power and piped water outages throughout Texas. Flores et al. (2020a) documented disparities in physical and mental health and health care access among Houston-area residents following Hurricane Harvey, prompting them to prioritize the need to ameliorate public health disparities resulting from climate change–related disasters. The team has published extensively on the intricate relationship between social vulnerability and harms to population health in the Houston metro area (Flores et al., 2020b; Griego et al., 2020), leading to adverse event experiences that prolong recovery.

Health System Performance

Compounding these high rates of disease and risk behavior are limited access to health care and social services due to factors such as income, systemic racism and discrimination, and geographic location. These limitations delay treatment and increase the risk of poor health outcomes. Not only are GOM populations vulnerable to illness and chronic health conditions but their health systems are also vulnerable. Composite rankings for 2019–2021 across multiple measures of "health system performance"—including health care access, quality, use of services, costs, health disparities, reproductive care and women's health, and overall health—rank GOM states in the lowest 30 percent, with most in the bottom 20 percent: Alabama (42nd), Louisiana (43rd), Texas (48th), and Mississippi (51st), lowest in the nation (Commonwealth Fund, 2023). Another indicator of health systems performance is an adequate health care workforce, assessed by Health Professional Shortage Areas (HPSAs). HPSAs are defined as "geographic areas, populations, or facilities" that "have a shortage of primary, dental, or mental health care providers" (HRSA, 2023). HPSA calculates scores that identify areas of greatest priority in need of improved health care services and workforce availability, of which the GOM, particularly Texas, Florida, and Mississippi, have a sizeable amount (Rural Health Information Hub, 2024).

During and oftentimes in the aftermath of a disaster, it becomes increasingly challenging to access social services, including services such as housing, food, and education provided by government, and private, for-profit, and nonprofit organizations for the benefit of the community and to promote social well-being (NASEM, 2019b). Health care services are also limited and people may deprioritize seeking care due to more pressing needs such as food, water, and/or shelter. Barriers also include insurance coverage and the availability of and access to culturally appropriate, high-quality care, including preventive, primary, specialist, dental, and vision care; chronic disease management; mental health treatment; and emergency services (NASEM, 2022c). These barriers can lead to increased morbidity and mortality rates. Similarly, people lacking access to mental health services or experiencing social isolation may face severe mental health crises.

Disaster Mental Health

The psychological footprint of a disaster is larger than its medical footprint—more people are affected psychologically than medically—because of the compelling nature of the event and the network of connections that extends geographically and socially beyond the scene of the event (Shultz, Espinel, Galea et al., 2007; Shultz et al., 2017). Populations exposed to disasters are known to experience elevated risks for psychological distress and psychopathology, including not only post-traumatic stress disorder (PTSD) but also other common mental disorders, including major depression, generalized anxiety disorder, and panic disorder, as well as increases in alcohol dependence and substance use (Pietrzak et al., 2012; Shultz et al., 2016). Further, preexisting mental health conditions can be exacerbated by the stress and trauma caused by a disaster, and even by the prospect of a disaster. Psychological consequences occur on a spectrum of severity, from distress to detrimental behavior change to diagnosable mental disorders, generally in relation to the severity of exposure to a hazard when it occurs and hardships in its aftermath (Davidson and McFarlane, 2006).

It is well known that disasters result in high levels of stress, which is possibly the most persistent, harmful, and universal adverse health consequence of disasters. Allostatic load, which is a concept used to physiologically measure the cumulative burden of chronic stress and life events or the wear and tear on the body, contains elements that can help elucidate increasing exposure to disaster-induced stressors (Guidi et al., 2021; McEwen and Stellar, 1993; Rodriguez et al., 2019; Sandifer et al., 2022). Because pervasive toxic stress as a result of disasters can lead to adverse health effects in the aftermath of the event, Sandifer and colleagues (2022) developed a framework that combines aspects of psychosocial and physiological allostatic load to estimate its burden in people who have experienced disasters and other traumatic events. The framework can be used to gauge the short- and/or long-term health effects of disasters and to predict and mitigate related illnesses or conditions, and can be deployed as a disaster-focused human health–observing system.

Psychological consequences are strongly dependent on the defining features of a disaster to which a person is exposed (i.e., the specific hazards, harms, losses, and changes, or combination thereof associated with the disaster; Shultz and Neria, 2013) as well as the collective consequences of and response to the disaster (Kirmayer et al., 2010).

Psychological consequences also extend over a prolonged duration. When a disruptive event occurs, people may have traumatic experiences that for some may progress to PTSD. Disaster-related PTSD is unique in that it occurs in large population groups at once, often overwhelming community mental health care systems (Espinel, Kossin et al., 2019). Long after the event has occurred, psychological consequences can persist as a result of ongoing hardships; losses (of resources and of loved ones, which can lead to traumatic bereavement); life changes (e.g., displacement); and risks for depression and anxiety disorders (Davidson and McFarlane, 2006) and collective trauma. Collective trauma can undermine a collective sense of security and can extend into second and third generations of survivors (Hirschberger, 2018). It is described by sociologist Kai Erikson as a "blow to the basic tissues of social life that damages the bonds attaching people together and impairs the prevailing sense of community" (Erikson, 1976, p. 153). GOM residents, one study (Lorenzini et al., 2024, p. 3) notes, "experienced a cascade of collective traumas in recent years."

Current and ongoing explorations of compounding disasters provide opportunities to advance understanding of the field of disaster mental/behavioral health, a discipline that has evolved rapidly over recent decades. The GOM region has been a primary locale for exploration of how compounding disasters affect mental health. Studies have shown how frontline health professionals experienced exhaustion as they dealt with their own storm recovery while deploying to the sites of multiple other storms (Herberman Mash et al., 2013), and how successive hurricanes severely affected their sleep patterns and provoked hyperarousal responses (McKibben et al., 2010). Fullerton and colleagues (2013) found that 8 percent of the statewide public health workforce experienced new-onset, hurricane-related PTSD or depression, along with associated increases in both smoking and alcohol use. A dose-response relationship was evident. The likelihood of PTSD and depression increased with each additional storm that directly affected a worker's home community and with each additional storm to which the worker was deployed.

Seminal disaster mental health research derives from studies conducted on GOM storms, most notably starting with Hurricane Katrina. Galea and colleagues (2007) explored exposures to hurricane-related stressors during Katrina in relation to mental health outcomes. Of particular concern, treatments were disrupted for people with preexisting mental illness (Wang et al., 2008).

In 2008, Hurricane Ike moved across the GOM, aiming directly for Galveston Island, and residents were strongly advised to evacuate to the mainland. Some did not, and they experienced the full force of the storm. Tracy and colleagues (2011) examined new-onset PTSD and depression in survivors of Hurricane Ike. Describing PTSD as "a disorder of event exposure," the researchers found that those most likely to develop hurricane-related PTSD had stayed on Galveston Island, experiencing Ike's wind and flooding hazards directly. The subset of Ike survivors who presented with clinically significant PTSD self-reported experiencing life threat ("I thought I was going to die"), losing a family member or close friend, or being physically injured in the storm. Stressors for depression included displacement from the home for a week or more, severe home damage, and significant financial hardship.

Sophisticated geospatial analyses identified clusters of individuals with hurricane-related psychopathology, concentrated in particular on Galveston Island in the aftermath of Ike (Gruebner, Lowe, Tracy, Cerdá et al., 2016; Gruebner, Lowe, Tracy, Joshi et al., 2016). Cerdá and colleagues (2013) followed Hurricane Ike survivors to examine "post-disaster mental health" after the event in relation to the stressors experienced. Lowe et al. (2013); Lowe, Fink et al. (2015); Lowe, Joshi et al. (2015); and Lowe et al. (2016) conducted a series of analyses that also looked at poststorm stressors and use of mental health services in relation to measures of both common mental disorder symptoms and wellness indicators. These studies of survivors of GOM storms were pivotal in teasing apart the effects of direct traumatic exposure as distinguishing predictors of those who would be diagnosed with PTSD as a result of experiencing loss and life change, diagnoses that were closely tied to diagnoses of depression, generalized anxiety, and substance use. Climate change is predicted to increase the frequency and severity of hurricane-related mental disorders (Espinel, Galea et al., 2019; Espinel, Kossin et al., 2019).

In 2010 the *Deepwater Horizon* disaster was a technological disaster involving release of a hazardous material (oil) on an expansive scale. Unlike a hurricane, this disaster posed more unknown and uncertain risks to residents. GOM researchers immediately began to document its effects on mental health. Osofsky and colleagues (2011) were among those at the forefront of this research, examining negative mental health effects for populations most heavily exposed. A concerning finding was that symptoms of depression in women increased over multiple time points, although symptoms of distress diminished (Rung et al., 2019).

Given the unique and unexpected nature of the widespread population exposures due to the *Deepwater Horizon* disaster, Shultz and colleagues (2015, p. 58) conducted a "trauma signature analysis" of the event, with "psychological risk characteristics of this event includ[ing] human causation featuring corporate culpability, large spill volume, protracted duration, coastal contamination from petroleum products, severe ecological damage, disruption of Gulf Coast industries and tourism, and extensive media coverage."

Studies of residents in the GOM region have shown links between exposure to severe hurricanes and PTSD, psychological stress (Cohen et al., 2023; Raker et al., 2019) and its comorbidities, substance abuse, and suicide (Gradus et al., 2010; Ouimette and Read, 2014).

Recent experiences with Hurricane Harvey became a focal point for research on the mental health of disaster-exposed populations. Karaye, Ross et al. (2019) and Karaye et al. (2020) examined self-rated mental health for Texas GOM residents who were exposed to Hurricane Harvey compared with persons without hurricane exposure. They found that GOM residents have poorer self-rated physical and mental health than the overall U.S. population. Analyses also showed the close interrelationship between repeated exposures to hurricanes and adverse effects on both mental and physical health. When comparing hurricane-exposed and nonexposed samples, those who had experienced a hurricane exhibited poorer self-rated mental health. Poorer mental health was reported by women, younger adults, mobile home residents, and persons with lower levels of educational attainment. Authors indicated that the self-assessment findings support the "enhanced provision" of mental health services coupled with educational programs and economic supports. Using the same 12-item Short Form (SF-12) Health Survey as Karaye, Ross et al. (2019), Sansom and colleagues (2021) found progressively lower scores on mental health on the SF-12 in relation to increasing numbers of direct hazard exposures over the past 5 years.

Relatedly, an international team of investigators (Li et al., 2021) conducted exploratory analyses on the role of neighborhood green space on coping with mental distress—and mitigating PTSD—following exposure to Hurricane Harvey. Perceived higher quality of green space, mediated through emotional resilience, was associated with lower PTSD.

Grineski et al. (2022) and Grineski, Scott et al. (2023) examined PTSD symptoms (anxiety and depression) in relation to the "cascading disasters" of Winter Storm Uri during the peak of the COVID-19 pandemic. A poststorm preliminary analysis of the data (Grineski et al., 2022) found an incidence rate of 18 percent. Uri-associated, new-onset PTSD was more common in minority race/ethnicity respondents than in White, non-Hispanic respondents.

Guided by a "cascading disaster health inequities" approach, Grineski, Scott et al. (2023) examined anxiety (using GAD-2) and depression (using PHQ-2)[7] in eight Texas metropolitan areas 6 months after Winter Storm Uri. Experiencing more adverse events and living with a disability increased odds of depression and anxiety. The detailed analyses present a more nuanced picture. First, minority racial/ethnic status was associated with greater odds of depression, but not anxiety. Second, a series of "doubly impacted" subsets—those who were Black and disabled, Hispanic and disabled, disabled and Uri-impacted, Black and Uri-impacted, or Hispanic and Uri-impacted—experienced elevated odds of depression compared with the applicable "doubly privileged" reference category.

COVID-19 Impacts on Frontline Responders

A number of studies were conducted that examined the mental health impacts of working on the frontlines during the COVID-19 pandemic. As a globally declared pandemic and a U.S. national public health emergency, first responders, hospital-based "first receivers," and public health preparedness coordinators were called upon to lead aspects of the COVID-19 response.

Mendez and Horney (2023) conducted qualitative interviews to examine the nature of the COVID-19 response from the vantage of Emergency Medical Services personnel in Texas. The pandemic set in motion a novel set of personal and professional stressors that included fear of transmission to friends and family, increased workloads, operational changes, and fatigue, along with impediments associated with their usual repertoire of coping

---

[7] GAD-2 is an initial screening tool to identify symptoms of "Generalized Anxiety Disorder." PHQ-2 or "Patient Health Questionnaire-2" is an initial screening tool to identify symptoms of depression.

skills. Authors indicate that there is a need to prioritize and implement evidence-based interventions to safeguard the health and well-being of frontline response personnel.

Pfender and colleagues (2022) conducted survey research to explore the dynamics of anxiety and depression among public health workers charged with responding to COVID-19. In a pandemic, it is the public health workforce that operates the frontlines. Investigators noted the dearth of knowledge about the prevalence of anxiety and depression within the public health emergency preparedness workforce and possible applications of social support to safeguard these workers. This cross-sectional survey revealed high rates of anxiety (40 percent) and depression (29 percent), and symptoms and diagnoses associated with burnout and suicide among frontline personnel. These findings require rapid implementation of robust supports at both the individual and organizational levels.

Medically High-Risk Patients

Medically high-risk patients (MHRPs)—those whose health conditions require them to have ready access to health services, systems, and often social services—face elevated risks from hazards (Balbus and Malina, 2009; Espinel et al., 2022; Heagele and Pacquiao, 2019; Shultz et al., 2024). They include people who overtly qualify as having a disability, such as those living with spinal cord injury (Shapiro et al., 2020), and whose health conditions are characterized by functional impairments (Kruger et al., 2018; Mitra et al., 2022). Medically high-risk patients also include individuals with chronic diseases who generally do not define themselves as disabled but whose medical conditions require a regular, dependable connection to critical medical treatments, such as cancer therapies (Nogueira et al., 2020; Shultz et al., 2024). MHRPs also include individuals whose special medical needs are transient but risk elevating in times of disaster. Examples include pregnant persons and those in postsurgical recovery. The physical, psychological, and socioeconomic challenges associated with living with chronic medical conditions can make it particularly difficult for medically high-risk patients to cope with the added stressors experienced in a disaster (Espinel et al., 2023). **Medically high-risk patients are at disproportionate risk for physical harm, psychological distress, and disruption of their care systems during disasters and other extreme events. They are also at elevated risk for such environmental hazards as extreme heat, humidity, heavy precipitation, and pollution.**

Since access to health services and supplies is frequently disrupted when a disaster occurs (Adams et al., 2019; Alnajar et al., 2021; Balbus and Malina, 2009; Espinel et al., 2022; Kruger et al., 2018; Mitra et al., 2022; Shultz et al., 2024), medically high-risk patients are likely to experience interruption of care routines, aggravation of symptoms, increased health care needs while evacuating and sheltering, increased susceptibility to injury, increased risk of cardiorespiratory events, and elevated stress levels. For example, exposure to hurricane hazards was associated with higher mortality among patients with lung cancer whose radiotherapy treatments were disrupted (Nogueira et al., 2019) and patients with end-stage kidney disease whose hemodialysis treatments were delayed (Blum et al., 2022), compared with patients who were not exposed. Yet although medically high-risk patients have special needs during a disaster, they tend to be underprepared compared with their healthy, disease-free counterparts (Adams et al., 2019). **The specialized needs of medically high-risk patients in the face of compounding disasters are inadequately incorporated in disaster planning, preparation, and policies.**

On a broader scale, the subset of medically high-risk patients with chronic and life-threatening diseases who must be safeguarded during a disaster is particularly concentrated in the GOM region (Achenbach et al., 2023). Even in the absence of compounding disasters, life expectancies in the GOM region declined steeply during the COVID-19 pandemic. Life expectancies have been extremely slow to rebound postpandemic (Achenbach and Keating, 2023), and markedly so in the GOM region. This means, in part, that medically high-risk patients in the region are at particular risk due to (1) the burden of chronic, life-threatening disease; (2) continuing disproportionate risk of severe illness and death from COVID-19; and (3) elevated risk during compounding disasters when the health care and social services and support systems, on which so many with chronic diseases rely for their survival, are disrupted.

Specific to the GOM region, Chakraborty and colleagues published a series of papers outlining elevated risks for "people with disabilities" during Hurricane Harvey (Chakraborty, Grineski, and Collins, 2019), Winter Storm Uri (Chakraborty et al., 2023), and the COVID-19 pandemic (Chakraborty et al., 2024). Chakraborty, Grineski, and Collins (2019) make a series of compelling points. First, people living with disabilities are vulnerable to disasters, yet environmental justice studies have generally not focused on this subpopulation of MHRPs. Second, disabled populations were overrepresented in Hurricane Harvey–flooded areas, and people with cognitive and mobility impairments were especially at heightened risk. Third, the authors advocate for inclusion of persons with disabilities in disaster planning and environmental justice research.

During the massive power outages that accompanied Winter Storm Uri (Chakraborty et al., 2023), people with disabilities, and most notably those living in federally assisted rental housing, sustained more severe and prolonged utility service disruptions, colder temperatures, and delayed recovery compared with nondisabled people—experiences that jeopardized their health. Authors call for the formulation of remedies that provide equitable protections.

Detailed analyses (Chakraborty et al., 2024) revealed marked disparities in adverse impacts of the COVID-19 pandemic by disability status in metropolitan Texas in five areas of life: mental health, physical health, living conditions, health care access, and social life. Additional disparities were uncovered when analyses were disaggregated by disability type. Respondents experiencing cognitive and independent living difficulties were the most disadvantaged by COVID-19 in all five areas of life. Authors once again emphasize the need for increasing research and for enacting disability-inclusive policies to protect MHRPs.

Social Capital and Cohesion

The multidimensional concept of *social capital* can be defined as "features of social organization, such as networks, norms, and trust, that facilitate coordination and cooperation for mutual benefit" (Putnam, 1993, p. 2). Social capital evolves through relationships between people, and between people and institutions (e.g., decision-making bodies, elected officials, emergency managers, nonprofit organizations) and is linked to the size and type of an individual's personal networks as well as other types of capital possessed by individuals in those networks, (i.e., financial, educational, cultural, and experiential resources; Bourdieu, 1985; Elliott et al., 2010). These relationships are central to a well-functioning and resilient society (NASEM, 2021). Within communities, social capital constitutes the social connections formed between and among community members (Putnam, 1993; Putnam and Goss, 1995). Social capital is sustained through social memory (Adger et al., 2005; Colten et al., 2012) and activated via social and civic networks (Aldrich, 2012; Putnam, 1993). These networks can help individuals cope with stress and trauma both in daily life and in times of a disaster. Given the likelihood for increased compounding climate hazards, broader support, including financial (Elliott et al., 2010) and mental health services (Ritchie and Long, 2021), is needed if social capital is going to be effective for all populations in times of disasters to build adaptive capacity.

Social capital also fortifies health at the individual level through access to resources and information and functional support (e.g., borrowing money; Clay and Abramson, 2021). Research has documented the linkages among compounding climate hazards, social capital, and community health resilience (i.e., the ability of a community to withstand, adapt to, and recover from public health risks; Roque et al., 2022).

Social cohesion is "the glue that bonds people to one another, in families, groups, organizations, and communities" (Rodin, 2014, pp. 61–62). It builds shared values, reduces community disparities, and can help communities thrive by fostering well-being through safety, social inclusion, self-determination, and preservation of cultural and traditional ceremonies. Being able to physically gather together is an important aspect of social capital and cohesion, as well as a major source of community health resilience and recovery in the wake of a disruptive event. Compounding disruptive events can make gathering together, and hence social cohesion and access to local social support networks, far more difficult by reducing physical connectedness; they can also disrupt intergenerational knowledge transfer through increased outmigration (Roque et al., 2022).

Social capital and cohesion, while intangible and difficult to quantify, are critical in all phases of a disaster (Broderick, 2023). These connections and networks are the infrastructure within which trusted relationships and solidarity among community members, or social cohesion, are nurtured (NASEM, 2024).

During and after a disruptive event, many people report turning to their social networks for assistance (Adger et al., 2005; Aldrich, 2012) in well-being and livelihood support regarding safety, social inclusion, self-determination, and maintenance of traditional ceremonies. These networks warn their neighbors of impending disruptive events; lend assistance in their wake; and assist with rebuilding even in the absence of formal, government-sponsored assistance (Airriess et al., 2008; Aldrich and Meyer, 2015; Colten et al., 2012). They enable informal arrangements to direct mutual aid, foster creative problem solving, deliver trusted risk communications, and facilitate swift responses to emerging challenges. This is a common, long-standing response in GOM communities (Colten et al., 2015).

The role of social cohesion in disaster phases has received attention, but its influence on the speed of postdisaster recovery remains understudied

(Bergstrand and Mayer, 2020; Elliot et al., 2010; Sobhaninia, 2023). One recent study (Sobhaninia, 2023) attempts to compare social cohesion and recovery rates in four different Puerto Rican communities affected by Hurricane Maria (in 2017), which was experienced by more than 3.4 million people. More than 200,000 individuals were displaced or left their homes, potentially impairing their social networks. With multiple hurricanes impacting the island since 2017, research showed that social cohesion plays a large role immediately after a disaster and provides many benefits, but that over time, its effect in advancing disaster recovery diminishes. Social cohesion may also be a double-edged sword in certain instances, with some pointing to the two-faced nature of certain types of social capital (Aldrich, 2012; Aldrich and Meyer, 2015; Howell and Elliott, 2019; Priest and Elliot, 2023; Smiley et al., 2018). The close ties of social capital can also perpetuate collective trauma and prolong recovery (Priest, 2023) as people tend to experience disasters not as isolated individuals but as members of localized social networks (Priest and Elliott, 2023).

As discussed in this chapter, disasters have a profound effect on the mental health of community members. In the aftermath of a disaster, social capital and cohesion can become even more important as community members deal with acute individual and collective stress during disaster response and recovery. Social capital can also provide social and material resources that shield against mental health stressors associated with both chronic and acute disruptive events (Torres and Casey, 2017). People with weaker support systems, such as those who are socially isolated or lack a close-knit community, may experience particular struggles with emotional and mental health challenges.

Neighborhood and Built Environment

Codes and standards are intended to minimize the vulnerability of buildings and infrastructure, but they face challenges in practice. First, these requirements are generally not retroactive, meaning that older structures, even if well maintained, do not comply with the latest technical guidance on hazard-resistant design, including evolving hazard characteristics. Because the U.S. building inventory is slow to turn over and building code reform remains a contentious political issue, the majority of buildings and other infrastructure are not designed to the latest construction standards.

The second challenge is the evolutionary nature of hazards, which is particularly problematic in an era of climate change. **Codes and standards are statistically regressive in nature, reflecting an attempt to predict the demands posed by various climatological stressors based on historical data. This means that buildings and other infrastructure designed to the latest technical guidance do not anticipate how the demands of even routine hazards (e.g., tropical storms in the GOM) may intensify over the lifetime of those structures.** Furthermore, codes may not consider or may discount "gray swan" events—those with significant consequences but low probability of occurrence. It is unclear, in fact, that science presently has the ability to project future hazards reliably (Williams, 2023). This pervasive vulnerability to an evolving hazard landscape is only compounded by issues unique to how specific building classes and infrastructure is designed, built, and maintained.

Public Infrastructure

The vulnerability of public infrastructure has been well established by the assessment and advocacy efforts of the American Society of Civil Engineers, which publishes national- and state-level Infrastructure Report Cards, presented for five GOM states—Alabama, Florida, Louisiana, Mississippi, and Texas—in Table 2-3. In short, Alabama, Florida, and Texas have overall infrastructure grades in the C or "mediocre" range, with "general signs of deterioration" and "significant deficiencies in conditions and functionality, with increasing vulnerability" for some elements (ASCE, 2021a, 2021b, 2022b). Meanwhile, Louisiana and Mississippi are graded in the D or "poor, at risk" range, "with many elements approaching the end of their service life." In these GOM states, issues of "significant deterioration" and "condition and capacity" are prevalent, "with a strong risk of failure" (ASCE, 2017, 2020). The specific infrastructure sectors driving these grades, listed in Table 2-3, highlight glaring vulnerabilities in many of the systems intended to manage the use of, and mitigate risks associated with, water—a critical asset and hazard for GOM states.

These deficiencies are associated largely with underinvestment in the maintenance and replacement of aging infrastructure, particularly those elements that have deficiencies, have reached the end of their service life, or have capacities insufficient for current demands. Underinvestment by state and local authorities illustrates the direct link between poverty and physical

TABLE 2-3 Overall Infrastructure Grades from the American Society of Civil Engineers (ASCE), with Grades by Sector

Infrastructure Element	Alabama[a]	Florida[b]	Louisiana[c]	Mississsppi[d]	Texas[e]
Overall	**C-**	**C**	**D+**	**D+**	**C**
Bridges	C+	B	D+	D-	B-
Dams	Unknown	D-	C+	D	D+
Drinking water	C-	C	D-	D	C-
Energy	B	C+	N/A	C	B+
Inland waterways	D	N/A	D-	D	N/A
Levees	N/A	D+	C	D	D
Roads	C-	C+	D	D-	D+
Wastewater	D	C	C-	D	D

[a]ASCE, 2022b.
[b]ASCE, 2021a.
[c]ASCE, 2017.
[d]ASCE, 2020.
[e]ASCE, 2021b.

vulnerability.[8] Steady growth in population and industrial and manufacturing industries has only further strained the capacity of many aging GOM infrastructure systems and has left states such as Texas struggling to maintain the sheer volume of infrastructure its population demands. Box 2-4 provides an example of underinvestment in infrastructure in Texas and the vulnerability caused by such underinvestment.

GOM states rely on their natural waterways and coastlines for commerce, which leads to increased concentration of bridges. The high vulnerability of these infrastructure networks not only places local economies at

---

[8] A subtext of infrastructure vulnerabilities is "do more with less." Public infrastructure is funded by tax dollars. Many localities and states lack the revenue to complete the necessary inspection, maintenance, and upgrades. A recurring theme is the gas tax, used to maintain roads and bridges. These taxes have been frozen for decades in some states because of the low economic capacity of citizens, even though inflation has doubled the cost of doing business during that time. Meanwhile, demands are rising, with more people and more commerce in the GOM region during that same time, and now even greater demands from a changing climate. Governments are thus straining to do more with half the funding today, a situation that creates deferred maintenance and infrastructure that is well past its design life, posing safety threats and increasing vulnerability to future hazards.

BOX 2-4
The Texas Power Grid and Winter Storm Uri

In 2021, the vulnerability of the Texas power grid came into stark relief during the blackouts that occurred during and after Winter Storm Uri. While Uri's extreme cold temperatures were uncommon in Texas, they are not entirely unprecedented. Estimates show that a 1989 cold snap in Texas would have exceeded the per capita energy demand of the 2021 blackouts (Doss-Gollin et al., 2021). Extreme cold events in Texas in 1951, 1962, and 1983 are also estimated to have resulted in about 90 percent of the per capita energy demand for heat as the 2021 blackout. These examples highlight the need for a power grid that is more resilient to extreme weather events (Doss-Gollin et al., 2021). (For additional discussion on the realized impacts of this widespread power grid failure in Texas, including the death toll, see Chapter 3.)

---

risk but also heightens the vulnerability of neighboring communities to floods and isolation. Finally, the public wastewater system vulnerabilities across the GOM region are notable, creating a significant threat to public health, as well as the potential for environmental impacts. Private water systems can also have vulnerabilities, as private wells and septic systems are at increased risk of microbial contamination during disasters (Gitter et al., 2023; Pieper et al., 2021).

Building Inventory

Building inventory provides a critical complement to public infrastructure in enabling economic activity; facilitating the delivery of education, health care, and government services; and housing the population. Like public infrastructure, building inventory (publicly and privately owned) has pervasive vulnerabilities driven by deferred maintenance and replacement of older buildings and the incompatibility of existing buildings with evolving hazard demands. Unlike infrastructure, however, buildings shelter and protect human life, and most are privately owned, creating additional challenges in addressing vulnerabilities to damage. The primary mechanism for doing so at scale—building codes—is based on a life safety approach aimed at maintaining affordability at the expense of potentially substantial losses from a disruptive event. Facilities such as housing are not intended for sheltering in place or being rapidly reoccupied after a disaster in which they are likely to sustain damage. Thus, while lives may not be lost, particularly if populations evacuate promptly, the physical damage to these buildings will disrupt functionality, displace populations, strain insurance and recovery assistance programs, and exert a significant financial and emotional toll on building owners.

The potential for mounting losses is only heightened in the GOM region by inconsistent adoption and enforcement of the latest model building codes. As noted in the Insurance Institute for Business and Home Safety (IBHS, 2018) report *Rating the States: 2018. An Assessment of Residential Building Code and Enforcement Systems for Life Safety and Property Protection in Hurricane-Prone Regions, Atlantic and Gulf Coast States*, while Florida has led the nation in advancement of building codes to limit hurricane losses, earning the top score of 95/100 in their rating system, other GOM states such as Texas, Alabama, and Mississippi were among the lowest-rated states nationally immediately preceding 2020–2021, largely as a result of the lack of adoption and enforcement of mandatory codes statewide. Even when adopted, codes are generally not regressive, governing only new construction and major renovations. Thus, while Louisiana made great strides in adopting and enforcing the latest building codes after Hurricanes Rita and Katrina, the building inventory is slow to turn over, so the vast majority of existing buildings—including those exposed to the weather-climate events of 2020–2021—were likely not constructed in accordance with modern building codes. This is particularly significant for GOM states, where population and housing booms during 1950–1960 resulted in what is now an older housing inventory.

Table 2-4 summarizes the percentage of housing units likely built to modern building codes in five GOM states. Florida is a recognized leader in the early adoption and enforcement of a statewide building code, leading to nearly 30 percent of its housing inventory likely built to model codes. Louisiana started enforcing modern codes after Hurricanes Katrina and Rita, so a larger proportion of its inventory is built to modern standards, but that proportion is still less than 5 percent. In contrast, Texas has undergone rapid growth in recent decades, resulting in over 13 percent of its housing inventory being built after 2010, when modern codes began to penetrate coastal and metro areas despite the lack of statewide mandated code adoption and enforcement. Note that Alabama does not enforce

TABLE 2-4 Percentage of Housing Units Built in a Single Year and in the Modern Code Era in Alabama, Florida, Louisiana, Mississippi, and Texas

	Alabama[a] (%)	Florida[b] (%)	Louisiana[c] (%)	Mississippi[d] (%)	Texas[e] (%)
Built in a Single Year (2020–2021)	2.25	2.90	1.91	1.66	3.66
Likely Built to International Building Code/ International Residential Code (IBC/IRC)	8.24	29.96	4.67	0.00	13.08

[a] Alabama does not have a mandatory statewide building code system. Since state codes reference the 2015 IRC for voluntary adoption, and coastal counties and major metropolitan areas were likely building to such standards, the census reference category "built 2010–2019" through "2020 or later" are adopted therein, assuming half the properties the first term met the new standards.

[b] The Florida Building Code, or FBC, was enacted in 2002 and draws upon national model building codes and national consensus standards, with modifications for the Floridian context. Therefore, the census reference categories "built 2000–2009" through "2020 or later" are adopted therein, assuming 80 percent of the properties the first term met the new standards.

[c] The Louisiana State Uniform Construction Code, or LSUCC, referencing the IRC became effective in 2007 with some exclusions and modifications; thus, the census reference categories "built 2000–2009" through "2020 or later" are adopted therein, assuming 20 percent of the properties the first term met the new standards.

[d] Mississippi adopted a building code law in 2014 that governs construction of most residential buildings, but with opt-out provisions. As a result, adoption and enforcement are currently uneven across the state.

[e] In 2001, the Texas legislature adopted the 2000 IRC as the standard for residential construction; however, the state does not mandate its adoption and enforcement. Given the variable uptake and delays in enforcement in coastal and metro areas that have adopted modern codes, census reference categories "built 2010–2019" through "2020 or later" are adopted therein.

SOURCE: U.S. Census Bureau, 2022b.

statewide building codes but references the International Residential Code, so estimates in Table 2-4 should be viewed with caution; the situation is even more dire in Mississippi, where no homes could be reliably assumed to be built to modern codes.

Housing

While the underlying vulnerabilities discussed above affect the entire building inventory, they are particularly salient for housing. Housing is an oft-neglected though central pillar of societal infrastructure. Stable, long-term housing is a foundational element for physical, social, and economic well-being, both individual and collective (NASEM, 2018), and can provide a sense of belonging in a community. Housing has also taken on new roles since the pandemic, enabling productive activities such as e-learning, telehealth, and telework to continue during lockdown periods (Mendonça et al., 2023)—activities that were reinstated when schools, businesses, and health care facilities were physically damaged in disasters. Thus, access to affordable and hazard-resistant housing is essential to the existence, flourishing, and resilience of communities.

Beyond the aforementioned issues of an aging housing inventory, the nation was plagued by a crisis in affordable housing well before 2020, which was exacerbated in the GOM region by the extreme weather-climate disasters and pandemic that subsequently occurred. Research suggests that disasters "disproportionately affect poor households living in exposed areas, often leading to displacement" (Norwegian Refugee Council, 2019, p. 41). This phenomenon is driven by a lack of safe, affordable housing options that forces lower-income populations into older or deferred-maintenance properties that are more likely to sustain losses even in more frequent, lower-intensity events such as Hurricane Sally or Winter Storm Uri. **The intersection of socially vulnerable households living in physically vulnerable housing creates vast segments of the GOM population who are especially ill equipped to recover from compounding disasters.**

Despite its vital role, housing is treated like a commodity: nonengineered, privately developed, minimally regulated after its initial construction, and seen by many as a pathway to building wealth rather than core infrastructure that all residents have a right to access. However, physical vulnerabilities within the built environment and the infrastructure supporting it (e.g., water networks) are the primary drivers of material losses in disasters.

Although housing recovery is certainly challenging for all disaster survivors who sustained housing damages or loss altogether, communities of color residing in lower-income areas tend to sustain more damage and recover more slowly compared with those in higher-income areas (Lee and Van Zandt, 2018; Peacock et al., 2014; Smiley et al., 2018). Moreover, while it is difficult to quantify the loss of cultural identity or cultural traditions important to populations, recovery processes typically overlook and undervalue historical significance and community cultural identities in favor of rebuilding newer, up-to-code, or safer alternatives for wealthier individuals. The damage to housing stock resulting from a disaster opens up new opportunities for redevelopment or "climate gentrification" (NASEM, 2022c), which often excludes or "outprices" generational community residents.

The physical loss of housing after a disaster drives up the cost of housing, rent, insurance, and property taxes. It also can contribute to loss of mental health stability, safety, sense of and attachment to place, and community cohesion, affecting all sectors of society.

Homeownership and a legacy of discriminatory housing practices   In the GOM region, as in other parts of the United States, the legacy of discriminatory federal housing policies and practices relegated many Black Americans and other minority residents to neighborhoods and regions with disproportionate exposure to climate and industrial hazards and environmental contaminants. Practices broadly known as redlining,[9] which was not outlawed until the Fair Housing Act (42 U.S.C. 3601 et seq.) was passed in 1968, included racial (or exclusionary) zoning[10] where, as early as 1910, local governments in GOM states with large Black populations (and in many other U.S. cities) instituted racial zoning ordinances to preserve segregation where it existed (Gray, 2022; Seicshnaydre et al., 2018), restricting where Black and other potential homebuyers of color could purchase homes. The exclusion of opportunities for federal housing loans[11]

---

[9] Prior to 1968, the Federal Housing Administration (FHA) did not insure mortgages in and near low-income urban neighborhoods where the vast majority of urban Black Americans lived (Fishback et al., 2022). Concurrently, the FHA subsidized construction of subdivisions for White Americans with the requirement that none of the houses be sold to African Americans (Rothstein, 2017).

[10] Prior to the Supreme Court's 1917 *Buchanan v. Warley* decision, city zoning ordinances across the United States legally forbade minorities from residing on blocks where a majority of residents were White (Rigsby, 2016).

[11] The Federal Housing Administration, established in 1934, enacted federal loan underwriting guidelines that explicitly supported racially homogeneous, suburban developments, effectively subsidizing loans for White homeowners, while systematically excluding racially mixed or predominantly minority neighborhoods that largely prevented homeownership opportunities and access to federally insured mortgages to non-White residents and expressly contributing to the concentration of poverty in urban areas (Taylor, 2019).

resulted in households of color receiving only 2 percent of all government-backed mortgages between 1934 and 1962 (Swope and Hernández, 2019).

These discriminatory practices resulted in long-term disinvestment and stigmatization (Keene and Blankenship, 2023) of these regions and neighborhoods that have residual effects into the present day, including residence in areas more prone to disruptive events, entrenched racial segregation, generational poverty, an inability to reside in safer areas, and overall housing affordability challenges, which disproportionately affects communities of color, compounding existing inequities in wealth and opportunity (Taylor, 2019).

In combination with existing patterns of segregation, redlining has resulted in "indelible patterns of social and environmental inequalities" (Grove et al., 2018, p. 524) with sweeping implications for health outcomes, including higher rates of cancer, asthma, poor mental health, and people lacking health insurance (Nardone et al., 2021) as well as residence in high-risk floodplains and more physically vulnerable structures (Bidadian et al., 2024) and reduced present-day greenspace (Grove et al., 2018), which all can be exacerbated in compounding disasters. The disruptive events of 2020–2021 largely reinforced the persistently unequal access to home-ownership (Joint Center for Housing Studies, 2021).

Displacement from home   Disaster-related housing shortages prompt both short- and long-term population displacement. As populations are displaced, the loss of a viable tax base can remove economic revenue that further erodes communities lacking intentional investment in recovery.

When disasters destroy subsidized housing in disaster-prone areas, the failure to rebuild leaves displaced residents without viable options for recovery beyond relocation (Mehta et al., 2020). For example, the loss of government-subsidized housing replaced with condominium developments eliminates the possibility of recovery in that area for those who are displaced, contributing to community erosion and a new normal that is not representative of the historical significance of what has been lost. Whether the negative effects of redlining and flooding, inequitable insurance practices, inflation of property values, or policies and processes that result in loss of property, displacement and relocation are significantly more common among vulnerable communities.

Large-scale housing losses are often accompanied by the absence of (or a significant lag in) access to temporary housing close to the household's place of work or children's schools, forcing low-income families to choose among a set of undesirable options that include remaining in an otherwise uninhabitable damaged home, making onerous daily commutes, permanently relocating, living in a vehicle or overcrowded housing, or experiencing homelessness. Following the disruptions associated with Hurricane Katrina, assessments revealed that one-third of displaced children were at least one grade behind in school for their age (Baussan, 2015). For more information on well-being outcomes related to Hurricane Katrina, the Resilience in Survivors of Katrina, or RISK, Project, a longitudinal, multi-method, multidisciplinary study, focuses on "the recovery of people, rather than place." The project has published more than 40 studies by following vulnerable individuals and their families if/when they relocated from New Orleans after the storm (Waters, 2016). Yzermans and colleagues (2005) note that psychological symptoms persist for more than 2 years after a disaster, and the displaced are twice as likely to have such symptoms.

Prolonged or permanent displacement, such as that created by experiencing compounding disasters, has significant impacts on security, mental health, social connectedness, well-being, and more. Disasters interrupt society, and in some instances, cultures. Long-term displacement ranks among the highest of the societal costs of disasters. Baussan (2015) notes that "evidence indicates that the climate displaced, particularly those who are low income, can suffer from greater hardships than they did prior to evacuation" resulting from challenges with housing, employment, education, and health (p. 2). Among the many losses experienced by these populations is the loss of a sense of and attachment to place, as discussed in Box 2-5.

Renters and rental homes   For renters, high demand for and low supply of housing are accompanied by a rising cost burden; between 2017 and 2022, rental prices increased at a rate faster than that of inflation (Schaeffer, 2022). Furthermore, between 2019 and 2021, the affordable housing shortage in the United States increased, with the number of affordable and available rental homes for extremely low-income renters[12] decreasing by more than 500,000 units, or 8 percent. Two GOM states, Texas and Florida, were among the top five states facing the most severe shortages (National Low Income Housing Coalition, 2023). Mississippi, while among the five states

---

[12] Households with incomes at or below the federal poverty guideline or 30 percent of American Median Income, whichever is higher (National Low Income Housing Coalition, 2023).

BOX 2-5
Sense of and Attachment to Place

The concept of place can be understood as a structure that includes both human experiences and the material world in which those experiences occur, or, as Casey (1997, p. 9) describes it: "an embodied experience—the site of a powerful fusion of self, space and time." Place is related to physical and mental functions or "nested collections of human experience" (Hess et al., 2008, p. 468) that engender intimate and personal knowledge of living in a specific place. Identities are rooted in a sense of and attachment to place and are geographically connected to the places at risk (Adger et al., 2011) in the GOM region. As the physical landscape changes in a place (e.g., hurricane-damaged coastline), peoples' identities can be deeply affected (Burley, 2010; Simms, 2017). Sense of place and place attachment are terms used to depict how people create, organize, and relate to places where social networks are rooted and both material and cultural flows coalesce.

Postdisaster, sense of place can play a key role in evaluating the meanings of long-term community recovery, as it did in a study (Schumann, 2018) with coastal Mississippi residents after Hurricane Katrina that compared quantitative and qualitative techniques. The author concludes that place-based qualitative assessments can enrich quantitative analyses and, together, achieve more equitable, holistic recovery appraisals.

Sense of and attachment to place can also assist residents in place (re)making after a disaster has changed their familiar physical surroundings (Zavar and Schumann, 2019). The dynamic nature of place attachment was highlighted in a study (Nelan and Schumann, 2018) that examined the interactions and limitations of informal and formal gathering places that emerged in the aftermath of Hurricane Harvey. Researchers found that, in part due to lack of formal mental health support resources, displacement of residents, and an influx of aid workers (many nonresidents), the gathering places were limited in their ability to foster a communal recovery among the residents.

---

with the lowest relative shortage of affordable and available units, had the highest concentration of renters (27 percent) in the United States who are behind in their housing payments and facing eviction in 2021. Louisiana is second with 25 percent (Joint Center for Housing Studies, 2021).

Renters are particularly vulnerable to disaster-related housing disruptions as a result of past and ongoing discriminatory housing practices, increased hazard exposure, physical vulnerability of structures, and socioeconomic disadvantage (Best et al., 2023). Weather-climate events tend to reinforce patterns of racial and class inequalities that were in place before a disruptive event, both in direct effects to housing and on household financial stress (Brennan et al., 2021; Lee and Van Zandt, 2018; Rumbach and Makarewicz, 2017).

Exclusionary practices and policies prioritize disaster aid for homeowners rather than renters, particularly wealthier ones. This has the effect of reinforcing discriminatory practices that lead to gentrification and isolation (Aune et al., 2020; Brand and Seidman, 2008; Ellen et al., 2016) and increases generational poverty, leaving communities with decreased capacity to recover from future disasters.

After a disaster, renters often have difficulty finding permanent housing near their original home because of rent increases, while the cost of newly rebuilt permanent housing often exceeds the ability of low-income disaster survivors to pay (Fothergill and Peek, 2004). This pattern also applies to residents of public housing development, who are also disproportionately affected (Chakraborty et al., 2021; Lee and Van Zandt, 2018). When public housing is destroyed in disasters, there is not typically a previously agreed-upon action for permanent reconstruction or recovery of the units (Hamideh and Rongerude, 2018). The former residents often find themselves with a reduced ability to participate in postdisaster decision-making due to displacement and a diminished number of residents able to advocate for reconstruction of government-subsidized housing units. Alongside the stigmatization of public housing and a lack of economic incentive to rebuild public housing (Hamideh and Rongerude, 2018), a failure to rebuild government-subsidized housing in disaster-prone areas removes viable options for recovery beyond relocation (Mehta et al., 2020). These conditions can exacerbate marginalization and leave residents more vulnerable to subsequent disruptive events.

Evictions increase after disasters and are associated with significant disruptions for renters in the years following a disruptive event (Brennan et al., 2021). Accelerated gentrification, challenges in accessing federal disaster recovery assistance, lack of clear title to the land, and overall lower supplies and higher demands are contributing factors to increased evictions postdisaster (Brennan et al., 2021; Schuessler et al., 2022). Rental units that remain tend to be older, poorly maintained, and not built to the latest building codes. Weather-climate hazards such as hurricanes, tornadoes, or winter storms will disproportionately damage such subpar properties.

Manufactured housing units   Manufactured housing units (MHUs) and MHU parks are different from rented or owned houses in ways that can render the units and their inhabitants more vulnerable to climate-weather disruptive events. Not only are MHUs typically flood-exposed at higher rates than other housing (Tate et al., 2021), they are especially vulnerable to high winds (Gabriel, 2023). MHU parks are also largely divided-tenure communities wherein residents own their housing unit, but rent the land beneath, a setup that restricts the rights of park residents (Rumbach et al., 2020).

The socioeconomics of the GOM region and the lack of affordable site-built housing have created increasing reliance on manufactured housing units. MHU parks, where many MHUs are situated, are the largest unsubsidized source of affordable housing in the United States (Sullivan et al., 2021). The populations residing in MHUs are more likely to have higher social vulnerability than the general population (Fothergill and Peek, 2004).

Of the 200,162 new MHUs shipped to states in 2020 and 2021, 77,242, or 39 percent, were shipped to the five GOM states (U.S. Census Bureau, 2023a). The six counties/parishes that are a focus of this study (with the exception of Harris County) all have a higher concentration of MHUs compared with the average for the continental United States (Strader, 2023). Cameron Parish, Louisiana, has the highest concentration (29.7 percent in 2000, 38.1 percent in 2018), 7 times the average of the continental United States. Of all MHU parks across all six parishes/counties, 23.6 percent are within special flood hazard areas;[13] highest percentages among these are in Cameron Parish (71.4 percent) and Galveston County (47.5 percent).

---

[13] Special flood hazard areas (SFHAs) are areas that "will be inundated by a flood event having a 1 percent chance of being equaled or exceeded in any given year (FEMA, n.d.). The 1 percent annual chance flood is also referred to as the base flood or 100-year flood. Furthermore, "lenders must require flood insurance on improved real estate or manufactured homes that are, or will be, located in a SFHA in communities that participate in the National Flood Insurance Program. [Federally] regulated lenders are prohibited from making, increasing, extending, or renewing any designated loan unless the real estate or manufactured home securing the loan is covered by flood insurance for the term of the loan" (Tobin and Calfee, 2005, p. 43).

Insurance   As discussed in Javeline et al. (2021), mounting catastrophic risk has prompted the insurance industry to not only raise premiums and deductibles, but eventually to deny coverage and avoid underwriting policies in Atlantic and Gulf Coast states (Hartwig and Wilkinson, 2016). Residual markets have naturally emerged in response to these market pressures, e.g., Fair Access to Insurance Requirements (FAIR) Plans and Beach and Windstorm Plans (Kousky, 2011; Randlett, 2010), but their explosive growth in recent decades (Hartwig and Wilkinson, 2016) reveals that they have now become the primary insurance market rather than the "market of last resort" in many hurricane-prone states (Laird et al., 2017; Randlett, 2010). When coupled with the heavy regulation of premiums, the shift into these residual markets has left these plans highly vulnerable to a catastrophic hurricane, with the majority of the plans already recording at least one underwriting loss (Hartwig and Wilkinson, 2016), with similar concerns over sustainability expressed regarding the National Flood Insurance Program (Ahmadiania et al., 2019; Arnold, 2011, pp. 235–236; Craig, 2019). Critiques further argue that these government-subsidized wind/hurricane and flood plans are actually working in opposition to their stated goal of reducing losses in these storms (Craig, 2019; Union of Concerned Scientists, 2014).

While the growth of residual insurance markets as the primary insurer in coastal zones is a clear signal of escalating risk, these states continue the process of "adverse selection" (Rowell and Connelly, 2012; Wagner, 2019) by pooling insurers into coastal underwriting associations required to insure the highest risk properties, even repetitive loss properties, at actuarially low premiums inconsistent with their actual levels of risk (Randlett, 2010). The consequences in the GOM region have been significant. For example, Louisiana insurers paid out more than $23 billion during the historic 2020 and 2021 hurricane seasons, causing 11 major insurers to declare insolvency (Barry, 2023; Finch, 2022). Meanwhile, the uninsured and underinsured often look to FEMA assistance, which is not only insufficient and slow to reach survivors (Kousky et al., 2021) but has similarly been depleted by excessive payouts in recent years (Kimball, 2023).

Economic Stability

Economic stability ensures that individuals and families have access to financial resources, job security, healthy foods, safe housing, and health care, all factors that improve quality of life and health outcomes and reduce vulnerability to disruptive events.

Median household income provides a metric for examining the distribution of wealth—and poverty. Median household incomes are generally low for families in GOM states. Mississippi has the lowest median household income in the nation, followed by Louisiana with the third lowest, and Alabama with the fifth. Among GOM states, only Texas has just above the average median household income for the nation (U.S. Census Bureau, 2021b).

GOM states also lead the nation in the percentages of their populations living below the federal poverty line. Louisiana leads the nation with the highest percentage of persons in poverty (19.6 percent), followed closely by Mississippi (19.4 percent); Alabama ranks fifth at 16.0 percent (U.S. Census, 2023b). The five GOM states are also home to 123 of the nation's 341 persistent poverty counties (counties with extended periods of high poverty rates). Mississippi has the highest percentage of any state (53.7 percent), encompassing 34.8 percent of the state's population (U.S. Census, 2023c). People who reside in persistent poverty counties experience more acute and systemic problems, including limited or lack of access to medical facilities and services, healthy and affordable food, quality education, and civic engagement opportunities (U.S. Census, 2023c), all important resources in the event of a disaster. GOM states have larger populations of racial/ethnic minority populations, lower aggregate incomes relative to the national average, and a higher percentage living in persistent poverty compared with non-Hispanic White populations in those states (Elder et al., 2007; Islam et al., 2022).

Disasters and Economic Instability and Inequity

Economic instability in GOM communities has cascading and compounding effects when disasters occur. During the response and recovery phases of a disaster, individuals may struggle to continue their employment because of the lack of paid time off and transportation challenges; low wages, a lack of savings, and insufficient insurance may result in their never recovering financially. This result in turn is a resource-loss spiral that leads to prolonged stress and poorer health outcomes (Sattler et al., 2002); research shows that poverty rates increase in the year following damage from natural hazards (Cutter et al., 2014; Gotham and Greenberg, 2014; Smiley et al., 2018).

Economic access to resources is one of the most significant factors related to inequitable recovery for frontline communities. Disaster recovery is arguably one of the best potential generators of improved outcomes given mitigation funding requirements, rebuilding processes, and insurance implications. However, the benefits of the mobilization of recovery funding rarely manifest among the vulnerable. Instead, disaster recovery processes often intensify vulnerabilities. These effects are further exacerbated by compounding and cascading events that place additional burdens on communities and continuously strain the capacity of struggling systems.

A longitudinal study (Smiley et al., 2018) examining the interaction of U.S. disasters (1998–2015), poverty, and various types of organizations involved in disaster recovery efforts found that poverty tended to increase the most where the number of "bonding," or more inwardly oriented organizations, increased. The robust, in-group solidarity these types of organizations rely on may unintentionally produce inequitable recoveries by restricting resources to group members while excluding members outside the group, particularly those most marginalized, and leaving them in comparatively worse situations than they may otherwise have been (Aldrich, 2012; Aldrich and Meyer, 2015; Howell and Elliott, 2019; Priest and Elliot, 2023; Smiley et al., 2018).

Generally, current disaster recovery approaches fail to address the root causes of social vulnerability and instability; thus, they often exacerbate preexisting vulnerabilities in tandem with losses from the disaster itself. Disaster-related inequality can be uniquely characterized by the unequal distribution of resources, power, and economic opportunity across societies in the United States (Bowdler and Harris, 2022). This diversion of resources is a key contributor to disparities seen across vulnerable communities.

Research shows that frontline communities, which are often, though not limited to, communities of color and low-income communities, are unlikely to benefit as much as their White, wealthier counterparts when qualifying for federal disaster assistance and home buyout programs (Billings et al., 2019; Emrich et al., 2022; Han et al., 2024; Loughran and Elliot, 2019; NPR, 2021). The Federal Emergency Management Agency also has reported that disaster funds are not reaching those most in need (FEMA, 2020). Of 4.8 million aid requests analyzed, the U.S. Government Accountability Office (U.S. GAO, 2020) found that

- the poorest renters were 23 percent less likely than higher-income renters to receive housing assistance;
- FEMA was roughly twice as likely to deny housing assistance to lower-income than to higher-income survivors because their damages were judged "insufficient"; and

- additional research indicates that people of color are less likely to receive adequate disaster assistance.

Following Hurricane Harvey, 40 percent of racial/ethnic minorities were denied assistance (Mosbergen, 2017). Additionally, 45 percent of households with annual incomes of less than $15,000 were denied individual assistance or received an average payout of approximately $4,300 (Texas Housers, 2018). Box 2-6 provides details on another barrier to receiving FEMA disaster assistance.

Policies such as redlining, deed covenants, heirs' property (see Box 2-7), flood zone designation, land-use planning and plans, credit score ratings, and discriminatory insurance practices have contributed to inequities. These inequities can be exacerbated during disaster recovery (Clark-Ginsberg et al., 2021; Domingue and Emrich, 2019; Finch et al., 2010; Finucane et al., 2020; Masozera et al., 2007; Whytlaw et al., 2021).

As growth in population and the built environment leads to greater exposure along coastlines at risk of experiencing extreme weather-climate events, economic instability, and other vulnerabilities to these high-impact hazard events is also increasing (Frazier et al., 2010). The convergence of repeated exposure to disruptive events and heightened vulnerability results in prolonged impacts and elevated sensitivity to future disruptive events (Smit and Wandel, 2006).

BOX 2-6
The 25 Percent Local Match Requirement

Access to Federal Emergency Management Agency (FEMA) public assistance funds requires that a financial match of 25 percent of the requested amount be provided by the requesting locality. Many smaller (by population) and more rural communities often have less capacity to navigate the funding application process and provide the dedicated "local match" required to qualify for the FEMA funding (U.S. GAO, 2022). Further, these capacity limitations constrain a community's ability to build the expertise needed to apply for funding that could help mitigate the effects of disasters and build adaptive capacity (Seong et al., 2021). Without adequate financial resources to support disaster recovery and mitigation, sensitivity to future disasters increases.

BOX 2-7
Heirs' Property:
A Significant Disaster Recovery Policy Change

Heirs' property is property passed to family members by inheritance, usually without a legal document proving ownership. It is typically created when land is passed on from someone who dies without a will to those legally entitled to their property, such as a spouse, children, or other relatives. These policies have contributed to the dispossession of 4.7 to 16 million acres over the last hundred years (USDA, 2023), disproportionately affecting African American families throughout southern states and accounting for an estimated one-third of all African American–owned land in the United States. The U.S. Department of Agriculture reported that "the lack of title severely limits property owners' ability to access credit, to sell natural resources, or to participate in land improvement programs offered by the Federal Government, resulting in land and wealth loss for affected families across the South," adding that it contributed to a 90 percent decline in Black-owned farmland nationwide between 1910 and 1997 (Gilbert et al., 2002; USDA, 2017, p. vii).

In Mississippi alone, the current value of the land lost in that period is roughly $6.6 billion (Deen, 2021). Until 2021, the Federal Emergency Management Agency's (FEMA's) policies required applicants for disaster aid to provide a deed or other formal proof of homeownership, which effectively excluded thousands of families seeking grants for repairs after disasters because of heirs' property restrictions (Dreier and Tran, 2021). However, new administrative policies announced by FEMA in 2022 offer flexibility in demonstrating proof of property ownership (FEMA, 2022).

SUMMARY OF KEY FINDINGS

The GOM region has a well-documented history of experiencing extreme weather-climate events, disruptions, and disasters. Scientific studies suggest that occurrence of rapidly intensifying and stronger tropical cyclones and bouts of extreme temperatures are increasing as a result of climate change. Shifting climate conditions will likely increase the frequency and type of hazards that GOM populations are exposed to, affording less time to recover before the occurrence of additional disruptive events. Exposure to extreme temperatures (heat and cold) are the deadliest weather phenomenon in the

United States, though these hazards receive insufficient dedicated federal support and funding for planning, education, mitigation, and recovery.

Evidence shows that many GOM residents experience high levels of social and physical vulnerability. GOM populations experience a range of poor health outcomes and health disparities and are marginalized in terms of income, education, age, ethnicity/race, political power, gender, sexuality, and/or being medically high risk. Many residents are disproportionately exposed to environmental contaminants and toxins, which exacerbate negative physical and mental health outcomes under blue-sky conditions and also increase risks to human health when hurricanes, flooding, and other weather-climate events mobilize these toxins into communities.

Vulnerabilities also extend to the built environment of communities. Codes and standards are statistically regressive in nature, reflecting an attempt to predict the demands posed by various climatological stressors based on historical data. This means that even buildings and other infrastructure designed to the latest technical guidance do not anticipate how the demands of routine hazards (e.g., tropical storms in the GOM) may intensify over the lifetime of those structures.

Vulnerabilities and exposure combine to produce a community's sensitivity to experience the occurrence of one or more disruptive events as a disaster that exceeds its capacity to absorb, respond, and recover. Communities with high vulnerability and exposure are more likely to experience long-term adverse effects of compounding disasters. Current disaster management and recovery practices fail to address the root causes of vulnerability and instability; thus, they often exacerbate preexisting vulnerabilities in tandem with the losses and damages directly associated with the disaster itself. Strengthening the adaptive capacity of the systems and functions that underpin communities can mitigate the potential for harm by reducing sensitivity to a disruptive event.

3

Compounding Disasters in the Gulf of Mexico Region: 2020–2021

This chapter begins with an overview of the disruptive weather-climate events of 2020–2021 in the Gulf of Mexico (GOM) region. Although these events occurred with notable frequency during this period, extreme and disruptive weather-climate events are not an unusual occurrence in the region. What sets the events in this time frame apart is their co-occurrence with a public health emergency—the COVID-19 pandemic. In tandem with the pandemic, each of the weather-climate events in 2020–2021 had consequences for the public's health in communities affected by these events.

The chapter then turns to focus on the content of the regional information-gathering sessions in Harris and Galveston Counties, Texas; Cameron and Calcasieu Parishes, Louisiana; and Baldwin and Mobile Counties, Alabama. This section of the chapter includes a description of the format of the information-gathering sessions, summaries of the content of the sessions, and discussion of the common themes that arose in each session.

SYNOPSIS OF DISRUPTIVE EVENTS

Detailing the costs, frequency, and number of previous disasters in the GOM region provides important context; however, to fully grasp the toll these compounding disasters exact on affected individuals and communities, a closer, qualitative examination is required. Informed directly by the voices of residents and practitioners working and living in the GOM region, this chapter details how disaster impacts compounded, intersecting with a complex set of sociopolitical, economic, cultural, infrastructure, and public health vulnerabilities among highly exposed and vulnerable populations, challenging the capacity of communities to recover.

The COVID-19 Pandemic

The COVID-19 pandemic has been intrinsically and inherently a public health event, both in form—a communicable disease—and in terms of leadership, with public health agencies and professionals, from the global to the local level, leading the response. COVID-19 spread expansively in time and space throughout the GOM region, complicating the dynamics of all other disaster events throughout the region. A graphic from the Fifth National Climate Assessment (USGCRP, 2023) illustrates how the overarching pandemic overlapped temporally with and was overlaid spatially upon other U.S. disasters throughout 2020 and 2021 (see Figure 3-1). Table 3-1 shows a comparison of the impacts on public/population health from 2020–2021 hurricanes determined to be billion-dollar disasters,

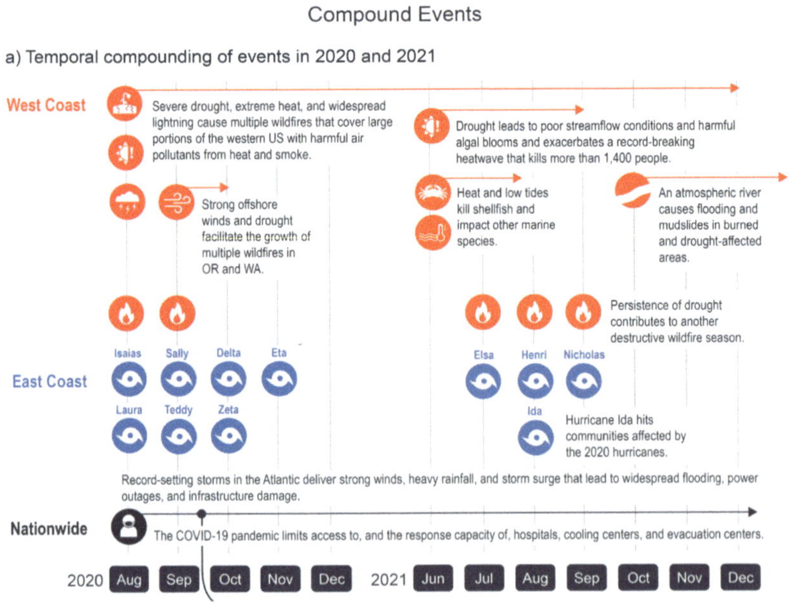

FIGURE 3-1 Temporal compounding of disruptive events in 2020–2021. The Gulf of Mexico region is included as part of the East Coast.
SOURCE: USGCRP, 2023.

TABLE 3-1 Profile of Disaster Impacts on Public/Population Health for Some 2020–2021 Disasters

Disaster Health Consequences		Hurricanes[a]	Winter Storm Uri	COVID-19
GOM Economic Costs ($)		1–75 billion[b]	200–300 billion[c]	Trillions[d]
Harms: Physical Health				
	Deaths	173[e]	246[f]	167,000[g]
	Injuries	Several thousand emergency room visits[h]	Several thousand emergency room visits[i]	Negligible
	Disease cases	Negligible	Hundreds (not related to COVID-19)	11 million confirmed[j]
	Medically high-risk patients	Increased risk	Increased risk	Life-threatening risk
Matrix: Psychological Stressors[k]				
	Potentially traumatic events	Exposure to storm hazards Personal injury Perceived threat to life	Exposure to storm hazards Personal injury Perceived threat to life	COVID-19 illness Hospitalization, ICU admission, mechanical ventilation Life threat Exposures at work
	Loss	Loss of life Damage/ loss of home, possessions	Loss of life Damage/loss of home	Loss of life Isolation Loss of employment
	Life change	Displacement from home Financial hardship	Displacement from home Financial hardship	Work/school disruptions Isolation Financial hardship Workforce stress Increased workload/ hours Mental exhaustion Balancing work/home demands

continued

TABLE 3-1 Continued

Disaster Health Consequences	Hurricanes[a]	Winter Storm Uri	COVID-19
Health Care Access and Disruption of Services			
Damage to facilities	Moderate to severe	Moderate to severe	None
Damage to power, infrastructure	Moderate to severe	Moderate to severe	None
Redirected services	Facility-specific—related to impact on facility Time-limited	Facility-specific—related to impact on facility Time-limited	Extensive—entire units/facilities shifted to COVID-19 Prolonged and periodic
Disruption of services	Facility-specific—related to impact on facility Time-limited	Facility-specific—related to impact on facility Time-limited	Extensive, widespread Prolonged and periodic
Public Health Services and Workforce			
Public health role	Active participatory role	Active participatory role	Lead role
Risk of harm/ death for public health personnel	Negligible	Negligible	Tens of thousands of public health professionals sickened and hundreds of frontline personnel died in GOM region
Disruption of services	Major but brief Community level	Major but brief Statewide in Texas	Extreme and prolonged throughout GOM region
Work stress for personnel	Significant	Severe/ relatively brief	Unprecedented/ prolonged Increased workload/hours Mental exhaustion Balancing work/home demands

TABLE 3-1 Continued

[a] Espinel, Galea et al., 2019; Espinel, Kossin et al., 2019; Shultz, Kossin, Hertelendy et al., 2020; Shultz et al., 2022.

[b] Xu et al., 2022.

[c] TASCE, 2022.

[d] CMS, n.d.; Hlávka and Rose, 2022.

[e] Xu et al., 2022.

[f] Texas Department of State Health Services, 2021.

[g] CDC, 2023c.

[h] AHRQ, 2020.

[i] Cindass et al., 2023; Orton et al., 2022.

[j] Johns Hopkins University, 2024.

[k] Shultz, Espinel, Galea et al., 2007; Shultz et al., 2013; Shultz et al., 2017.

Winter Storm Uri, and COVID-19; Table 3-2 shows total deaths, deaths per 100,000 GOM residents attributable to COVID-19, and rating GOM states for COVID-19 mortality.

COVID-19 Mortality in Gulf of Mexico States

COVID-19 was the third leading cause of death in the United States in 2020 and 2021. COVID-19 deaths dwarfed numbers of fatalities from all forms of disasters nationwide and in the GOM region. Four of the five GOM states ranked among the highest in COVID-19 morbidity rates in the nation in 2020, and three of the five ranked among the highest in 2021 (CDC, 2023c).

The racial disparities associated with COVID-19 were significant. In Louisiana, for example, Black Americans accounted for 50 percent of known COVID-19-related deaths while representing only 32 percent of the state's population (Hu et al., 2022). Links between disparate exposure to air pollution are predicted to be one of several factors contributing to the disproportionate deaths among Black Americans, racial/ethnic minorities, and other vulnerable groups (Chakraborty, 2021; Terrell and James, 2022).

COVID-19 Complications and Impediments for Hurricane Response

One of the most urgent problems posed by COVID-19 was how to safely evacuate and shelter populations in the path of hurricanes and tropical systems. Salas et al. (2020), Shultz, Kossin, Hertelendy et al.

TABLE 3-2 Total Deaths, Deaths per 100,000 Residents Attributable to COVID-19, and GOM State Rankings for COVID-19 Mortality Rates

	2020			2021		
	Deaths	Rate/ 100,000 Residents	State Rank for COVID-19 Mortality Rate	Deaths	Rate/ 100,000 Residents	State Rank for COVID-19 Mortality Rate
U.S.*	350,831	85.0	-	416,893	104.1	-
Louisiana	6,533	118.0	6	6,329	116.9	16
Mississippi	4,466	123.5	4	5,082	146.3	5
Texas	30,840	105.2	9	44516	151.4	3
Alabama	6,544	103.6	11	9,491	152.8	2

*U.S. data from Xu et al., 2022.
SOURCE: CDC, 2023c; Xu et al., 2022.

(2020), Shultz, Fugate, and Galea (2020), and Shultz, Kossin, Ali et al. (2020) have described the challenges and some clear incompatibilities that became evident in the early months of the pandemic. Specifically, Shultz, Kossin, Hertelendy et al. (2020, p. 494) characterize them as follows:

> Climate-driven hazards underscore the imperative for timely warning, evacuation, and sheltering of storm-threatened populations—proven life-saving protective measures that gather evacuees together inside durable, enclosed spaces when a hurricane approaches. Meanwhile, the rapid acquisition of scientific knowledge regarding how COVID-19 spreads has guided mass anti-contagion strategies, including lockdowns, sheltering at home, and physical distancing; life-saving strategies [that] separate and move people apart.

Gathering people *together* (e.g., in an evacuation shelter) for hurricane protection conflicts logistically with moving people *apart* to minimize risk of transmission of airborne communicable diseases. It is impossible to implement both approaches optimally and concurrently (Salas et al., 2020; Shultz, Fugate, and Galea, 2020; Shultz, Kossin, Hertelendy et al., 2020).

GOM states were among the first to ease restrictions and reopen their economies (Miller, 2020) after the lockdowns began in March 2020. Five months later, by mid-August 2020, GOM states experienced a massive COVID-19 case surge, leading the nation in newly diagnosed cases, hospitalizations, and deaths. Simultaneously, the 2020 Atlantic hurricane season produced a record total 30 named storms, including 11 that made landfall in the United States (NOAA, 2021m).[1]

In August 2021, the situation in the GOM region was repeated when, again, GOM states led the nation in a massive flare-up of COVID-19 cases. This time, however, it was the virulent Delta variant circulating—the most lethal variant during the course of the pandemic (Patrick et al., 2022).

COVID-19 Vaccination Rates in Gulf of Mexico States

Vaccines against COVID-19 were available by November 2021, but by then, Mississippi, Alabama, and Louisiana ranked first, second, and fourth, respectively, among states for the highest number of deaths per capita due to

---

[1] At the time of writing, the 2020 hurricane season continues to hold the title of the most active hurricane season on record.

COVID-19. Florida was not far behind at 9th, and Texas was 19th (Statista, 2021). At that same time, just 46.6 percent of Mississippians, 48.4 percent of Louisianans, and 45.6 percent of Alabamans were fully vaccinated (CDC, 2023d). These 3 states were among the 10 states with the lowest vaccination rates; their hospitalization rates were nearly 4 times higher and death rates more than 5.5 times higher than those 10 states with the highest vaccination rates (Hanna et al., 2021).

The low vaccination rates persisted through October 2022, when Louisiana, Mississippi, and Alabama, ranked 48th, 49th, and 50th, respectively, on the percentage of vaccinated 18- to 64-year-olds.

COVID-19 and Economic and Employment Stability

The COVID-19 pandemic significantly affected economic and employment stability nationwide (U.S. Census Bureau, 2024). Several years into the pandemic, incomes were lower overall and by state compared with 2019, just prior to the pandemic (U.S. Census Bureau, 2021c). Beginning in 2021, the sharp rise in inflation outstripped the rise in income, producing the net effect of decreasing buying power and the value of household savings, which were already low for GOM states (U.S. Bureau of Labor Statistics, 2023d). The nation's overall unemployment rate, which reached 14.8 percent during the pandemic, was extremely high (39.3 percent) for those working in the leisure and hospitality industries because of stay-at-home orders (Falk et al., 2020). Those who remained in the service, manufacturing, and transportation industries were required to report physically to work, putting them at higher risk of contracting COVID-19 because of increased exposure and lack of sufficient workplace safety measures.

Throughout 2020–2021, the emergence and spread of COVID-19 transformed the public health risk of compounding disasters in the GOM region by amplifying health-compromising exposures and underlying vulnerabilities and modifying emergency procedures for weather-climate events. Trivedi (2023, p. 1) highlights that, "while compounding disasters have long existed, the duration and scope of the COVID-19 pandemic both drew additional attention to them and the challenges of addressing them" (citing Tozier de la Poterie et al., 2022).

Hurricanes and Tropical Storms

In 2020–2021, three Category 3 or higher storms—Laura, Zeta, and Ida—made landfall in the GOM region (all in Louisiana). Additionally, Hurricane Sally (Category 2) made landfall in Alabama, and Hurricane Delta (Category 2) made landfall in Louisiana, and Hurricanes Hanna and Nicholas (both Category 1) made landfall in Texas. The events of these 2 years were situated in the context of long-term climate change and short-term variability associated with a "double-dip" La Niña (NOAA, 2021b). The GOM region's notable hurricanes and tropical storms are summarized in Table 3-3.

Other notable tropical systems impacting the GOM region in 2020–2021 include Hurricane Eta (in 2020) and the precursor storms associated with Tropical Storms Bertha (in 2020) and Fay (in 2020). Bertha ultimately affected the Carolinas, but flooding (more than a foot of rainfall) in South Florida caused by the evolving system resulted in structural damage in Miami and surrounding regions. A small tornado also affected the region. Tropical Storm Fred (in 2021) was a minor storm that made landfall in August on the western Florida panhandle, causing thousands of power outages. The storm ultimately caused $1.3 billion in damage and seven fatalities as it moved into the Northeast (NOAA, 2021k). The following month, another tropical storm (NOAA, 2021l) made landfall near St. Vincent Island, Florida. That storm spawned an EF0 tornado south of Tallahassee, Florida, and wind damage caused power outages and tree damage in much of North Florida. Figure 3-2 depicts the tracks of the hurricanes and tropical storms in the GOM region in 2020–2021.

Tornadoes

During 2020–2021, April 2020 was the most tornado-active month in the United States, seeing several multiday events (April 7–8, 12–13, 19, and 22–23; NOAA, 2020). Many of these outbreaks were in the GOM region (see Table 3-4). In 2021, tornado reports exceeded the 1991–2010 U.S. annual average (NCEI, 2022) with many of these events occurring in the GOM region (see Table 3-4), including an EF2 tornado to touch down in Lake Charles, Louisiana, in late October 2021, damaging about a dozen homes and injuring several people (Misick, 2021).

Note that there were multiple other wind and/or rain events that occurred in the GOM region in 2020–2021 that are not listed in Tables 3-3 and 3-4 but still caused loss and damages in the region.

TABLE 3-3 Summary of the Gulf of Mexico Region's Hurricanes and Tropical Storms 2020–2021

Hazard and Description	Landfall Date	Affected Areas	Monetary Damage in the United States ($)	Direct Fatalities
Hurricane Hanna Hanna's prehurricane impacts affected virtually the entire GOM region. Hanna's primary hazards included excessive rainfall, rip currents, and winds. New Orleans experienced street flooding as the outer bands from the developing storm skirted the Gulf coast (NOAA, 2021c).	7/23/20	S TX	1B+	5
Hurricane Laura Laura was a Category 4 storm that made landfall in Cameron Parish, Louisiana. Winds in excess of 150 mph, rainfall, and storm surge inundation levels of 12 to 18 ft above ground were described as "unsurvivable" and "catastrophic." Laura caused significant damage in coastal regions and inland cities, such as Lake Charles. Extensive housing and infrastructure damage included 568,000 power outages in Texas and Louisiana and an estimated 10,000 demolished homes in Louisiana. Shortly after the core of the hurricane passed Westlake, near Lake Charles, Laura triggered a fire at a chemical plant that discharged a cloud of potentially toxic smoke that stretched for miles. Laura was the strongest hurricane, in terms of maximum sustained wind speed at landfall, to affect Louisiana since 1856. The World Meteorological Organization retired the name Laura after this storm (NOAA, 2021d).	8/27/20	SW LA, SE TX, SW MS	19B	47

Hurricane Sally Sally was a Category 2 storm that made landfall near Gulf Shores, Alabama. Like many recent hurricanes (Harvey in 2017, Florence in 2018, and Dorian in 2019), Sally was a slow-moving tropical cyclone, producing significant flooding. While the storm featured 100+ mph winds, the rainfall totals in excess of 2 ft is a signature fingerprint of this storm (NOAA, 2021e).	9/16/20	AL, W FL and Panhandle, GA	7.3B	4
Tropical Storm Beta Tropical Storm Beta caused widespread flooding in the Houston area, disrupting major transportation corridors and prompting numerous water rescues (NOAA, 2021f).	9/17/20	SE TX, LA, MS, AL	225M	1
Hurricane Delta Winds and rainfall from Delta compromised already weakened structures damaged by Hurricane Laura. Impacts (structural damage and power loss) from Delta were felt in Louisiana and Mississippi (NOAA, 2021g).	10/9/20	LA, MS	2.9B+	2
Hurricane Zeta Hurricane Zeta attained "major hurricane" status as it made landfall in Cocodrie, Louisiana. It is the latest landfalling major hurricane on record for the continental United States. Zeta was a Category 3 storm and produced storm surge inundation of 6 to 10 ft above ground level along the Mississippi coast (particularly in the back bays) (NOAA, 2021h).	10/28/20	SE LA, Alabama, SW MS, N GA	4B	5

continued

TABLE 3-3 Continued

Hazard and Description	Landfall Date	Affected Areas	Monetary Damage in the United States ($)	Direct Fatalities
Hurricane Eta The final hurricane of the 2020 season, Hurricane Eta made three total landfalls, the latter two in Florida on November 9 and 12. Eta caused major flooding from storm surge and heavy rainfall, mainly in South Florida (NOAA, 2021i).	11/9/20	FL, GA, SC, NC	1.5B	7
Tropical Storm Claudette Claudette moved onshore south of Houma, Louisiana. Nearly a foot of rainfall caused damage in parts of Louisiana and Mississippi. Multiple tornadoes were spawned, including a long-track EF2* tornado in Alabama, where multiple fatalities were reported (NOAA, 2022d).	6/18/21	SE LA, MS, AL, GA, SC, NC	375M	4

Hurricane Ida Ida was a multihazard storm covering multiple geographic regions, with initial impacts in the GOM region. At Category 4, the hurricane made landfall close to Port Fourchon, Louisiana, with maximum sustained winds of at least 150 mph (240 km/h) and a minimum central pressure of 930 millibars. Ida was one of three hurricanes in recorded history to make landfall in Louisiana with 150 mph winds, along with Hurricane Laura in 2020 and the Last Island hurricane of 1856. According to the National Oceanic and Atmospheric Administration, 100 percent of homes in Grand Isle, a barrier island about 10 miles from the port, received damage; 40 percent were completely destroyed. There was heavy damage to the energy infrastructure across southern Louisiana, causing widespread, long-duration power outages that affected millions of people. Parts of New Orleans were without power for more than a week during the hottest months of the year. This timing amplified heat-related risks among an already vulnerable population. The World Meteorological Organization retired the name Ida after this storm (NOAA, 2022a).	8/29/21	LA, NE MS, NW AL, E and central and E TN, portions of KY and VA, SW VA	75B+	96
Hurricane Nicholas Nicholas made landfall in rural Matagorda County, Texas. Power outages (nearly 500,000 in Texas) and freshwater flooding were the primary outcomes in such places as Harris and Galveston counties. More than 100,000 Louisiana residents lost power during flood-producing rainfall, and Florida reported two fatalities due to rip currents (NOAA, 2021j).	9/12/21	SE TX, S LA	1B	2

*The Enhanced Fujita Scale (EF Scale) became operational on February 1, 2007. It is used to assign a tornado a rating based on estimated wind speeds and related damage. When tornado-related damage is surveyed, it is compared with a list of damage indicators, or Dis, and degrees of damage, or DoD, to help estimate the range of wind speeds likely produced by the tornado. From that information, a rating (EF0 to EF5) is assigned (National Weather Service, n.d.)

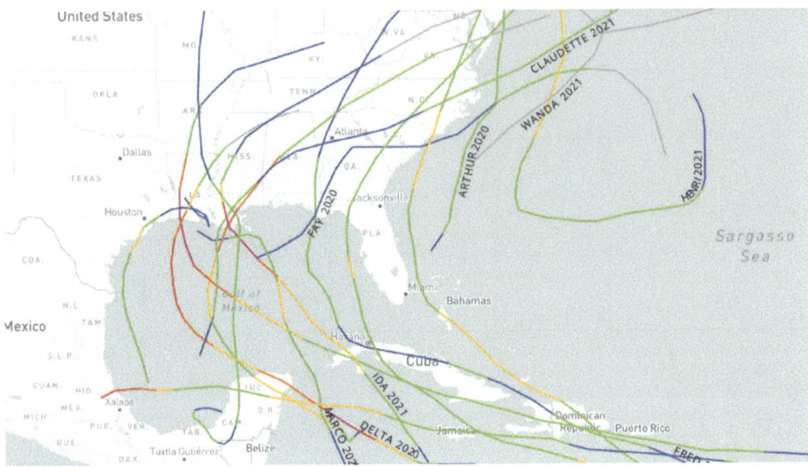

FIGURE 3-2 Tropical storm and hurricane tracks in the Gulf of Mexico region, 2020–2021.
SOURCE: NOAA, n.d.

Extreme Temperatures

Parts of the GOM region experienced extreme heat during 2020–2021. The impacts were amplified by episodic power losses associated with thunderstorms and hurricanes. The year 2020 was tied with 2016 for the warmest on record globally, and most of Florida endured record heat that year, even by its own rather warm standards (NOAA, 2021a). An early spring heatwave also broke temperature records as the result of a persistent high-pressure system over the GOM (Cappucci, 2020). Many people in the GOM region were exposed to these record-breaking heat events while enduring COVID-19 and episodic storms. During Hurricanes Delta, Zeta, Laura, and Ida, heat was the most frequent cause of deaths in Louisiana. Of the 65 deaths collectively attributed to the storms, 23 were due to extreme heat (Louisiana Department of Health, 2023).

Extreme cold temperatures and winter weather overwhelmed parts of Texas, Louisiana, and Mississippi during a weeklong storm event—Winter Storm Uri—in February 2021. This event is often mischaracterized as a polar vortex storm. The polar vortex is not a storm, but a circulating wind pattern (see Figure 3-3) that is always present above the Arctic region. When it is strong, extremely cold air is "trapped," but when it weakens (as it did in 2021), extremely cold Arctic air can spread south into the midlatitudes. This

TABLE 3-4 Tornadic Events in Gulf of Mexico States in 2020–2021

Date(s)	GOM State(s) Impacted
January 10–11, 2020	Mississippi, Alabama, and Louisiana
February 5–7, 2020	Mississippi and Alabama
March 18–19, 2020	Texas
March 30–31, 2020	Texas, Alabama, Mississippi, and Florida
April 12–13, 2020	Texas, Louisiana, Mississippi, Alabama, and Florida
April 19–20, 2020	Mississippi, Alabama, and Florida
April 22–23, 2020	Texas, Louisiana, Mississippi, and Florida
June 6–9, 2020	Florida
November 24–25, 2020	Texas and Mississippi
November 30, 2020	Florida
December 23–24, 2020	Texas
January 25–27, 2021	Alabama and Florida
February 15, 2021	Florida
March 13, 2021	Texas
March 16–18, 2021	Mississippi and Alabama
March 24–28, 2021	Texas and Alabama
April 19–20, 2021	Louisiana, Mississippi, and Florida
April 23–25, 2021	Texas and Alabama
April 27–28, 2021	Texas
May 2–4, 2021	Mississippi
December 29–30, 2021	Alabama

sequence of events enabled a strong Arctic cold front to plunge deep into the interior United States, bringing temperatures as much as 40°F (4.44°C) below normal throughout the region.

Within the past decade, significant advances have been made in the prediction of extreme weather-climate events (Sillmann et al., 2017). Yet, as is often seen with anomalous weather-climate events, forecasts for Uri provided a sufficient time window to initiate certain preparations. However, the urgency or awareness around preparations was lacking, demonstrating that **better preparation is needed at the individual, community, and regional levels for winter weather extremes, even if they are currently rare.**

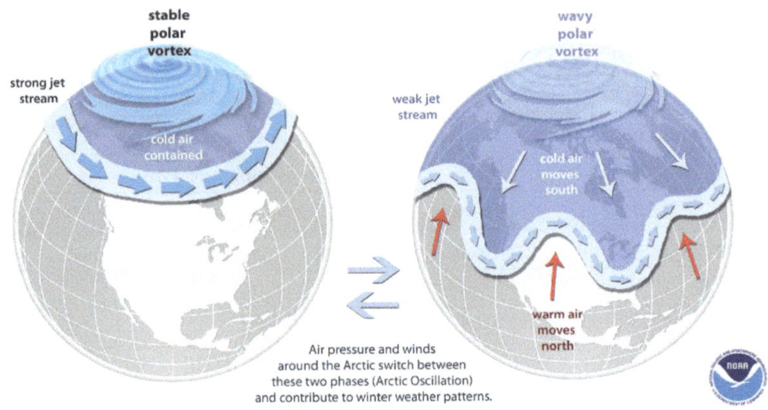

FIGURE 3-3 The science of the polar vortex.
SOURCE: NOAA, 2019.

IMPACTS OF COMPOUNDING DISASTERS IN REGIONS OF TEXAS, LOUISIANA, AND ALABAMA

To gain an understanding of how compounding disasters affect the lives and livelihoods of individuals who live and work in the GOM region, the Committee on Compounding Disasters in Gulf Coast Communities, 2020–2021, held a series of 13 in-person information-gathering sessions with local, state, and federal representatives from two adjacent counties/parishes in three GOM states: Alabama, Louisiana, and Texas. The primary data elicited from session participants in the information-gathering sessions formed the basis of the report's summary, analysis, findings, and conclusions. Potential participants were identified through their professional roles within disaster-affected communities, and invitations were extended to a wide range of invitees. In total, 77 individuals participated in the committee's in-person information-gathering sessions (see Appendix B for a list of panelists). Panelists who agreed to participate represented the subset of individuals who accepted the committee's invitation and were present for an information-gathering session. The committee is aware that the experiences shared at these sessions are only a small glimpse into the realities that existed for those affected by these events in the GOM region during this 2-year period, a time when everyday stresses were exacerbated by the anxiety and uncertainty brought on by the COVID-19 pandemic and multiple, successive weather-climate events.

Recovery Time Lines

The information-gathering sessions all began with a participatory activity aimed at documenting the disasters and recovery time lines in each region where that panel took place. The disaster sequences collectively captured in each session in a particular region were aggregated into a single time line, shown in Figure 3-4. Darker gray-shaded regions represent COVID-19 surges based on available case-count data, while yellow-shaded boxes mark the 6-month Atlantic hurricane season, illustrating that both the 2020 and 2021 hurricane seasons were accompanied by surges in COVID-19 cases, as noted earlier in this chapter. These time lines were populated by asking participants to first identify the disruptive events that were most significant for their community. For each of these events, they then collectively defined the recovery period that followed, represented by the colored (at times overlapping) bars in Figure 3-4. When characterizing the impacts and recovery time line for each of these events, respondents underscored that even when these events do not directly compound the losses from a preceding event, they continue to have indirect effects by interrupting and extending the ongoing recovery process (e.g., reconstruction and rebuilding of homes postdisaster will likely need to be paused in advance of an approaching storm).

Recovery Indicators

As part of disaster recovery, many organizations, residents, and state and federal agencies use indicators to measure response and recovery. When defining the recovery time lines in Figure 3-4, the committee asked session participants to identify specific indicators they view as signs that a community has "recovered" from a disaster. **The committee found that there is no single indicator or consistent set of indicators that represent when or if a community and its members have fully recovered.** The indicators cited most commonly by respondents were categorized based on six sectors that meet the critical needs of a community (see Table 3-5). Specific examples of indicators in each sector are detailed in the subsequent summaries of the information-gathering sessions.

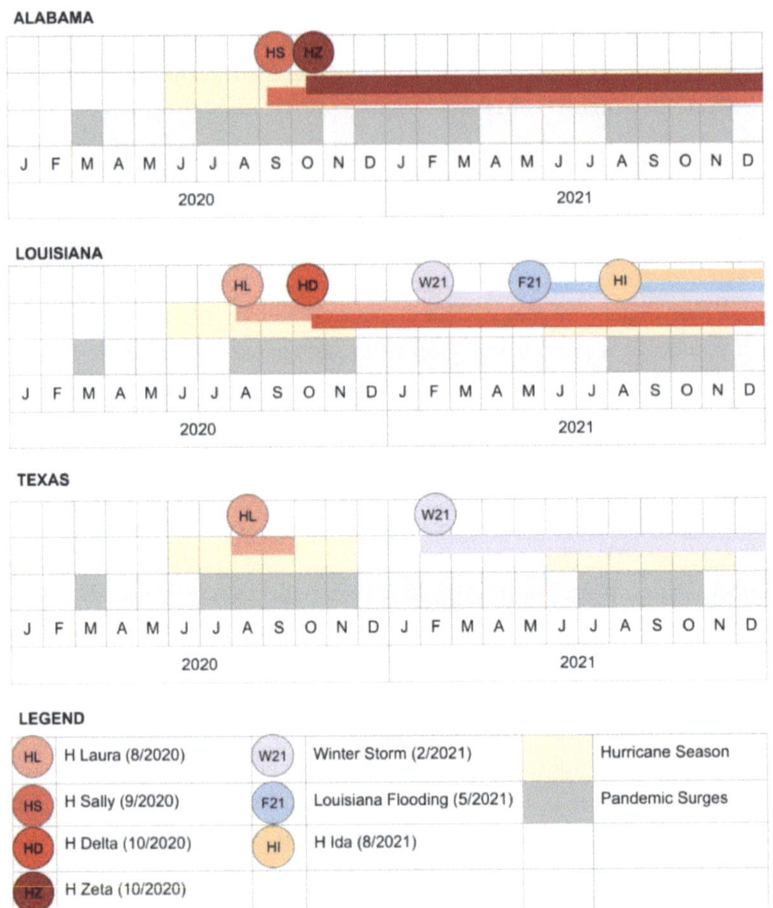

FIGURE 3-4 Aggregated recovery time lines for each state visited in the committee's information-gathering sessions, identifying specific disruptive events (circles) and recovery periods (colored bars) relative to pandemic surges (dark gray bars) during an ongoing pandemic period (light gray strip) and intermittent hurricane seasons (light yellow strip).

Following the definition of recovery indicators and time lines, discussions with participants focused on COVID-19's impacts on disaster preparation, adaptations and/or changes in preparations for the 2020 and 2021 hurricane seasons, specific impacts of the identified disasters, "bright spots," and lessons learned. The following sections summarize these discussions and identify key issues that surfaced across sessions.

TABLE 3-5 Most Frequently Self-Reported Recovery Indicators in Information-Gathering Sessions

Systems/Sectors	Common Recovery Indicators
Public Services	Utilities (water, power) and services (transportation) restored,* debris removed*
Local Economy	Businesses reopen,* tourism returns, entertainment resumes
Housing	Repairs complete,* population return,* building permits issued, tarps removed, insurance claims paid
Health Care	Mental health indicators,* services restored, wait times reduced
Education	Schools reopened*
Civil Society	Faith-based institutions reopened*

*Referenced by a majority of participants in a panel session.

HARRIS AND GALVESTON COUNTIES, TEXAS

On October 10 and 11, 2022, the committee held five 90-minute information-gathering sessions in Houston, Texas (Harris County), with representatives from community-based organizations (CBOs) and local, state, and federal government officials from Texas's Galveston and Harris Counties. On October 10, the committee held sessions with (1) leaders representing nonprofit and nongovernmental organizations (NGOs) involved in disaster risk management, housing, and community development (Harris NGO Panel); (2) public officials representing floodplain and emergency management, homeland security and public safety, housing, engineering, and public health in Harris County (Harris Officials Panel); and (3) state and federal officials, including a regional Federal Emergency Management Agency (FEMA) coordinator and representatives from the Texas General Land Office and the Small Business Association (Texas Federal/State Officials Panel). On October 11, session participants included (1) Galveston County nonprofit and NGO leaders representing response and recovery, education, financial stability, community emergency response, the Chamber of Commerce, and food security (Galveston NGO Panel); and (2) Galveston County officials representing public health, housing, and emergency management (Galveston Officials Panel). In total, 20 people participated in the committee's information-gathering sessions in Texas.

2020–2021 Disasters of Greatest Impact

In addition to discussing challenges brought on by the global COVID-19 pandemic, panelists from the Galveston and Harris County information-gathering sessions focused on Winter Storm Uri[2] as their primary hazard event. This extreme weather-climate event prompted large-scale power grid failures resulting in the inability to heat homes, burst pipes and water, disruption of internet and cell phone service, severe supply chain delays, and far-reaching economic hardship for residents and business owners unaccustomed to and thus unprepared for extended freezing temperatures (see Chapter 2 for more detail). Without power and heat, many businesses statewide, which were already hobbled by the pandemic, closed down for several weeks.

Panelists noted that community members were experiencing anxiety, malaise, and disaster fatigue at this time. One Galveston NGO leader working in disaster recovery was discussing ongoing disaster recovery projects from as far back as Hurricane Ike, when she stopped and said: "One of the things that we noticed, and you all have probably seen some level of it, a lot of our population, and some of our leadership, are really having disaster fatigue. When we were getting threatened with Nicholas, a lot of people went, you know what? To heck with it. I don't care. I'm exhausted. I'm exhausted from battling my way through my life in COVID. I'm exhausted from my pipes being broken and trying to get resources to get things fixed. I don't care if they knock my house down. . . . It's just been one trauma after another." Box 3-1 provides more detail on community disaster fatigue.

Recovery Indicators and Timing

Hurricane Laura, which made landfall in a parish adjacent to the Texas border, did not cause structural damage in Harris or Galveston County, but triggered traumatic recollections of Harvey's impacts. It also drew resources toward affected areas in southeast Texas and Louisiana and brought displaced Louisiana evacuees into the state. As a result, at the time of Winter Storm Uri, Texas was still sheltering Louisiana residents from Hurricane Laura, all within the restrictions necessary to manage the COVID-19 pandemic. A secondary hazard event, Hurricane Nicholas, a Category 1 storm that made landfall in mid-September 2021, also affected these areas.

The Harris and Galveston NGO panelists stressed that recovery from Uri was taking longer than hurricane recovery typically would, with

---

[2] For more information, see Donald (2021).

BOX 3-1
Community Disaster Fatigue

During 2020–2021, the combination of high-profile, high-impact disruptive events (e.g., weather-climate events and the pandemic) and racial inequality and police brutality contributed to *community disaster fatigue*, especially among low-income communities and communities of color. Ingham and colleagues (2023, p. 2) define community disaster fatigue as a "collective situation in which the components of a complex system fail to adapt appropriately, leading to inactivity, public health issues, and dysfunctional institutions." Community disaster fatigue can slow down or stymie much-needed changes in the health, economic, and societal arenas as the focus remains on the disaster recovery process (Ingham et al., 2022).

Outcomes of community disaster fatigue appear to be associated, on a collective level, with elements reflecting the breakdown of community resilience (Ingham et al., 2023). **With the recurrence of disasters, communities continue to struggle as disaster recovery becomes increasingly prolonged and unattainable. As a result, community resilience and adaptive capacity decline, compassion fatigue sets in, and key leaders and policymakers become complacent without delivering recognizable or meaningful change. Further, the struggles of daily survival create consistent stressors that impede effective healing in the midst of health, economic, and societal challenges.**

some people still unable to return to their homes. They informed the committee that recovery from prior hurricanes—Harvey (in 2017) and Ike (in 2008)—and floods—Tax Day (in 2016) and Memorial Day (in 2015)—was also still ongoing and was extended by Uri, particularly with respect to housing.

Panelists listed recovery indicators that included repair of burst pipes and backed-up sewer lines, restoration of power, a decline in boil-water notices, the ability to ship donated blood, hospitals being "normal busy" instead of in "catastrophic mode," and dialysis centers reopening with the ability to admit more patients. Harris County officials noted that disaster recovery is often seen at the community level and can be considered concluded when everyone is "whole" and back to work again. Meanwhile, panelists representing state and federal officials distinguished short-term recovery indicators such as restoration of utilities and resolution of transportation concerns, from long-term recovery indicators, such as finalizing of funding allocations and rebuilding or upgrading of infrastructure in accordance with new codes. One panelist mentioned the large number of homes typically listed in the market after a hurricane, defining recovery as the moment when these more-than-normal for-sale signs no longer appear. Other indicators that panelists identified included the restoration of access to adequate food and clean water, an increase in department staffing to predisaster levels, availability of housing for displaced populations, and shelters being closed.

The timing of recovery from Uri varied by community and neighborhood. Many roads took up to a week to reopen. Regarding power restoration, participants reported that many neighborhoods waited 10–14 days to have electricity restored, though as noted previously, housing recovery had not yet concluded by the time the information-gathering sessions were held.

Realized Impacts of 2020–2021 Disasters

Winter Storm Uri was a weeklong event that included multiple days of dangerously cold temperatures that affected the GOM region, compounding the severity of power and water outages, disruption of renewable energy production, and transportation stoppages. According to the National Oceanic and Atmospheric Administration (NOAA), nearly 10 million people were without power for extended periods, and studies have confirmed that a disproportionate number of those affected were vulnerable populations (NOAA, 2021o). Depending on the source, estimates of the death toll range from 246 (TDSHS, 2021) to more than 700 (Aldhous et al., 2021), with the *Houston Chronicle* contending that there were more than 1,000 unexplained deaths within the duration of Uri (Despart, 2022). As one NGO panelist said, "Nobody was prepared for Uri." The unprecedented scale of the Texas power grid failure made the storm response much more challenging, especially since power was necessary to maintain a safe level of warmth in residences.[3] The committee also heard about the extensive

---

[3] As discussed in Chapter 2, the death count for Winter Storm Uri varies depending on the data source. However, the most common cause of death is definitive: exposure to extreme temperatures. According to the Texas Department of State Health Services (2021), 65.4 percent of winter storm–related deaths were caused by extreme cold exposure-related injuries (hypothermia and frostbite). Hanchey et al. (2023) attribute only 47.1 percent of winter storm–related deaths to extreme cold exposure, but they still found that extreme cold exposure was the most common cause of death.

damage to infrastructure. Pipes froze, shifted, and broke, impeding water and sanitation services. The co-occurrence of Winter Storm Uri and the pandemic affected sectors at nearly every level. Hospitals were already overburdened before the power outages, which then further hindered their ability to operate effectively. One response organization resorted to using social media platforms to screen online posts in an effort to find people seeking help, explained one participant. Social media sites were also used to disseminate real-time information addressing community needs. Compared with Tropical Storm Beta and Hurricane Laura, "we did not see a lot of conversation . . . but when you come to the winter storm, it exploded. And it exploded for a little while, until peoples' phones lost their charge. And then we saw a decline in what people were posting, because they were sitting shivering in their house with no electricity to charge their phones," a participant explained.

The compounding effects of Uri impaired risk communication in general. Most session panelists noted that their southern tier geographies were unfamiliar with how to prepare for such a deep freeze, and many did not believe that the event's impacts would be so severe and sustained. Even with ample notice of the storm's arrival, many GOM communities did not anticipate being immobilized for an extended period of time. One NGO representative from Harris County explained, "Almost everybody has their hurricane season stuff ready. Nobody thinks about the other disasters that could potentially happen during the year. Being prepared for those is just as important." Box 3-2 provides more information as to cognitive bias challenges embedded in disaster understanding.

It was necessary to get disaster information out to residents, especially that pertaining to preparation for freezing weather, a generally uncommon event in the region. One NGO leader noted, "The emergency broadcast system—it lacks preparedness in letting people know what's happening. But if a child or something is abducted it flashes on everyone's phone. But no one was prepared for Uri. Well, that's because of the grid and everything. But we don't get those notifications in a timely manner. So, I think that needs to be addressed." Another state official expressed the need for greater collaboration with houses of worship and other local purveyors of information to improve pathways for trusted and accurate risk communication. Moreover, many people had relocated to Harris County prior to Winter Storm Uri, but many were unfamiliar with emergency protocols, shelter locations, and evacuation routes used in hurricane preparations that would have provided a level of familiarity with general emergency protocols.

BOX 3-2
Influence of Cognitive Biases on Risk Perception,
Planning, and Communication

When people confront the threat of an impending disruptive event or disaster, their decision-making can be impaired by the tendency to overemphasize recent events or personal experiences when estimating the impact of the future event (recency bias) and by the tendency to underestimate the likelihood of the occurrence or impact of the event (normalcy bias). **Historical events often serve as benchmarks for how people perceive and plan for future hazards. This tendency can be dangerous.** Recent examples of "benchmark" storms include Hurricanes Andrew (in 1992), Ivan (in 2004), Rita and Katrina (in 2005), and Gustav and Ike (in 2008) and the 2011 winter freeze. However, the problem with referring to storms as benchmark is that the bar is always shifting, and the potential for future events to surpass historical events is increasing. Hurricanes Harvey (in 2017) and Ida (in 2021) are examples of storms that render benchmark events obsolete. Accordingly, cognitive biases that inform how people interpret and respond to new threats can result in significant underestimation of risk, leading to inadequate planning and preparation, and ultimately to more severe impacts. The effects of cognitive bias are not limited to the decision-making of individuals; indeed, they may be reflected in regulations, codes, and policies.

According to traditional risk assessment, planners are advised to identify and prioritize risks with the highest probability of occurring locally. This approach can lead to the neglect of atypical and rarer events and leave communities susceptible to an unanticipated, but not unimaginable, event such as Winter Storm Uri. Responses to atypical events such as extreme cold can be outside the purview of organizations accustomed to response and recovery for hurricanes, tornadoes, and flooding, further complicating relief efforts. Those responding have steep learning curves in anticipating and responding to the needs, damages, and other impacts of such an atypical hazard.

Compounding events further complicate the traditional risk assessment methodology (Zscheischler et al., 2018).

For example, the rainfall intensities experienced now and going forward are not like those of the rainstorms of past centuries or even decades. The literature shows that the top 1–2 percent of rainfall-intensity events have increased in recent decades (Easterling et al., 2017). In just the last 15 years, the percentage of the southern United States experiencing extreme 1-day rainfall events in any individual spring has increased significantly (Di Liberto, 2021; USGCRP, 2023). Relatedly, jurisdictional infrastructure (e.g., roads, bridges) was designed and engineered under an assumption of "stationarity"—that is, assuming that hazards such as rainfall would have future intensities similar to those of rainstorms of decades past. A significant portion of U.S. infrastructure was designed and built in the 20th century. Not only is it aging and often in a state of deferred maintenance, but it now exists in the changing climate of the 21st century.

For centuries, European communities have placed physical markers to indicate the height of past floods (Glaser et al., 2010). Such public memorialization is one way to both document past events and provide an enduring reminder of past tragedies. Still, social memory can fade—within both professional circles and the public. Further, human recency and normalcy biases associated with disaster experience are vexing challenges. Programs dedicated to public and agency disaster risk communication require permanent funding, regular updating, community participation, and ongoing implementation.

Deficiencies and deficits in risk communication are also linked to vulnerability, particularly as related to a lack of comprehensive information, which can lead to false perceptions (Birkmann and Fernando, 2008) and influence motivation and perceived ability or need to prepare for disruptive events. To address this challenge, the National Hurricane Center (NHC) has been working with social scientists to improve communication products, notably the tropical cyclone "cone of uncertainty" product. In early 2024, the NHC announced the upcoming release of an experimental tropical cyclone forecast cone graphic for the 2024 hurricane season (NOAA, 2024).

The effects of Uri caused a tremendous loss of life, explained participants. People did not have heat, fuel was not readily available, and residents resorted to barbecuing in their garages for warmth. Many carbon monoxide detectors did not function properly, and people used outdoor propane heaters to provide heat in their homes. "It was devastating," summed up a panelist. "People have been traumatized by multiple successive disasters, and when another one strikes, it can feel like PTSD [post-traumatic stress disorder]," a Harris County NGO leader said.

Relatedly, participants noted a dramatic rise in domestic violence and socioeconomic impacts, resulting from or exacerbating mental health issues. Panelists from both counties experienced high turnover rates in the health care industry, and new staff were not always familiar with disasters or disaster response. This was but a microcosm of broader trends, with Houston's population declining for the first time after Uri. Businesses were also not returning at the typical rate, participants said. These ramifications had wide-ranging effects that, at the time of these panel sessions, were still ongoing.

CAMERON AND CALCASIEU PARISHES, LOUISIANA

The committee convened four information-gathering sessions in Lake Charles, Louisiana (Calcasieu Parish), on December 7 and 8, 2022, with representatives working and/or living in Louisiana's Calcasieu and Cameron Parishes and working in the disaster space. The opening session included leaders of CBOs, NGOs, and nonprofit organizations representing general aid and community support services, senior services, faith-based communities, housing rebuild/construction, disability advocacy, and health care (physiological and behavioral) (Louisiana NGO Panel). In the second session, the committee heard from Cameron Parish government officials representing floodplain and emergency management, health care, and parish administration (Cameron Public Officials Panel). The third session consisted of Calcasieu Parish government officials from parish administration, public health, homeland security and emergency preparedness, housing, and behavioral health (Calcasieu Public Officials Panel). The final session included state and federal officials representing FEMA, public health, community development, homeland security and emergency preparedness, mental and behavioral health, and emergency services (Louisiana State/Federal Panel). In total, 30 people participated in the committee's information-gathering sessions in Louisiana.

2020–2021 Disasters of Greatest Impact

In addition to the COVID-19 pandemic, panelists agreed that Hurricane Laura, a Category 4 hurricane that made landfall near Cameron, Louisiana, on August 29, 2020, 15 years to the day after Hurricane Katrina in 2005, was the most significant weather-related event to impact their region in the 2020–2021 time frame. Panelists' views differed as to the second most significant weather-related event because of the number of disruptive events occurring within a relatively short period. Hurricanes Delta and Zeta both hit in October 2020. Hurricane Delta, which mobilized debris and caused further damage to structures already damaged by Hurricane Laura, complicated recovery activities, as well as FEMA and insurance claims stemming from Hurricane Laura. Hurricane Zeta caused some flooding in Cameron Parish, and its impacts also siphoned resources to other areas, affecting Calcasieu and Cameron Parish residents still sheltering in New Orleans after Hurricane Laura.

Winter Storm Uri in February 2021 brought extreme cold weather with freezing rain, causing pipes to burst and acutely affecting water systems for homes and public buildings, such as hospitals and dialysis centers, that were still damaged from Hurricane Laura. Some residents had yet to return to their homes because of Hurricane Laura's impacts when Winter Storm Uri hit. Three months after Uri, a heavy May rainstorm brought about a foot of rain in less than 24 hours, flooding and reflooding areas in the region (Di Liberto, 2021) and further hindering ongoing recovery efforts. Panelists also made note of Hurricane Ida (in August 2021), clarifying that its direct impacts were not significant in this region, but that the storm drew resources and some attention away from Southwest Louisiana to the southeastern portion of the state; the added stress also compounded existing mental health challenges.

The impacts and respective recovery time lines of each event make it difficult, if not impossible, to differentiate between or ascribe specific impacts to each disaster. Many panelists also drew comparisons between Hurricane Laura and September 2005's Hurricane Rita. In a discussion about determining when an area can be considered "recovered," an NGO leader asked the group about recovery from earlier storms: "Did we recover from Rita honestly and thoroughly? . . . There is so much still from Hurricane Rita that is not repaired."

Recovery Indicators and Timing

Short-term recovery indicators, as reported by panelists, included the removal of blue tarps from roofs, schools reopening, rebuilding of the workforce (including having enough clinicians to meet the community's mental health care needs), and the return of residents and the entertainment industry. Long-term recovery indicators included the return and reopening of small businesses and schools, the availability of affordable housing and new construction, access to counseling resources, family stabilization, children achieving educational objectives, local communities meeting their needs without outside support, and the number of substance abuse cases and overdose deaths returning to predisaster levels. Local officials added to the list of recovery indicators the substantial completion of construction projects, utility restoration, repairs to damaged homes, and resumption of essential services, as well as removal of road debris and blight remediation. They mentioned as well the return to predisaster levels of alcohol sales, home assessment values, school enrollment, domestic violence, and divorce rates. State and federal officials cited additional measures: the ability to be proactive rather than reactive, a majority of building damage repaired, reopening of small businesses, and resumption of social and community activities. These discussions portrayed recovery as a phased process: the first stage concerns achieving stability, said a Louisiana State official, while the second includes incorporating resilience into future planning and actively making improvements.

Regarding recovery timing, all four Louisiana panels agreed that, for many residents of Calcasieu and Cameron Parishes, recovery from Hurricane Laura was still ongoing, particularly demolition, repairs, and restoration. Panelists noted that new construction was just starting at the time of the information-gathering session in late 2022. Panelists also agreed that COVID-19 and the series of climate and weather-related events following Hurricane Laura prolonged the overall recovery time line, prompting one panelist to quip, "The term compounding disasters—we've got to be the poster child for that." Having sufficient federal support and cooperation was mentioned as a key factor in the recovery process.

Realized Impacts of 2020–2021 Disasters

Session participants noted that the COVID-19 pandemic, Hurricane Laura, and the hurricanes and storms that followed affected all sectors of the local economy in Calcasieu and Cameron Parishes. Since the pandemic began, businesses of all kinds—including restaurants, hotels, casinos, and the entertainment industry—had been struggling to stay open. Many small businesses had still not been able to, or had chosen not to, reopen at the time of the panel session. If businesses were operating, most had reduced hours because of understaffing. As one panelist pointed out: "It's not good enough just to have a plan. You have a plan B, and a plan C, and a plan D."

Hurricane Laura's effects included widespread damage to public sector–operated utilities, with restoration of electricity taking 5–6 weeks. Session participants recalled how the main broadband provider, responsible for one of the few essential services provided by a private company, "failed miserably." One Calcasieu Parish official noted that perhaps the "biggest revelation for us was how deeply we had come to depend on broadband technology. . . . Overnight, businesses, schools, government, individuals [needed broadband] not just for entertainment but for basic services. . . . We realized we were very vulnerable." Those without access for longer periods struggled to access or learn about disaster resources, and it severely limited students' ability to attend virtual classes when their schools were damaged. The storm demolished radio towers, eliminating the ability to communicate hurricane response information. The Lake Charles radio tower still is not rebuilt, a panelist pointed out. Delta's winds caused even more power outages, in part because of trees damaged in Laura, further exacerbating utility recovery.

Calcasieu and Cameron Parish residents also struggled to access health care facilities after the storms. Hospitals were damaged and utility outages prevented adequate patient care. However, one of the complicating features of 2020–2021's compounding disasters was that, in some cases, readily available funding was earmarked for COVID-19 care. Yet, following a hurricane, the most urgent need was to reestablish care in area hospitals that were damaged by the storm. One panelist laid out the scenario, saying, "Almost every dollar that came down to the local health department was related to COVID [and] those dollars were very specific in what they could be used for. And there were bigger threats to people's health and well-being than COVID during this time period." (See Chapter 4, Box 4-1, for more information on COVID-19 federal funding.) For example, the only hospital in Cameron Parish was flooded with 2–3 feet of water and remains inoperative as of November 2023 (Maschke, 2023). This situation interfered with all types of hospital-based care and was especially problematic for medically high-risk patients whose medical conditions could not be treated adequately.

The education sector lost 6,000–7,000 students because of the 2020–2021 disasters, participants explained. As enrollment dropped, schools were allocated fewer resources or closed, and school budgets were reduced by about $50 million, said one panelist.

Multiple local government buildings and staff homes were damaged or worse, destroyed, which further hindered response. Law enforcement personnel faced similar challenges and lacked functional facilities or vehicles that could navigate the flooding and blocked roads. Fifteen of the 18 fire stations in Calcasieu Parish were so damaged that they could no longer store their fire trucks safely. Even funeral homes were damaged to the extent that families struggled to find facilities to bury or cremate their loved ones. Participants reported that every church in Cameron Parish was damaged, and as of December 2022, only two had reopened. Areas flooded that had never flooded before, in part because of accumulating hurricane debris clogging waterways, explained one participant. At one point, before Delta's landfall, residents were advised to "secure your debris," which they had heaped streetside following the instructions after Hurricane Laura. The call, though futile for most, was intended to prevent debris from remobilizing, whether by wind or water, and causing further damage. This instruction, one panelist said, is "the synopsis of compounding disasters." Delays and limited capacity in debris removal created obstructions to stormwater flow and contributed to flooding in areas that had never flooded before the May 2021 rainstorm.

Panelists in Louisiana emphasized the strong sense of solidarity and willingness to help each other during the compounding disasters of 2020–2021. They agreed that the concerted response to the weather-climate disasters reinvigorated a sense of community closeness and care for one another that COVID-19 had taken away. One panelist noted that it was a matter of survival: "Heading into a disaster where isolation was the name of the game, we were not going to survive as a community in isolation." The trust, rapport, collegiality, and genuine care expressed among individuals representing different governmental and nongovernmental actors was tangible in nearly all of the information-gathering discussions. "It takes a community," one federal official noted. "We all just needed to love each other," said an NGO panelist from Southwest Louisiana, "and that really, I think, brought us back as a community . . . because that is at the core of who we are."

BALDWIN AND MOBILE COUNTIES, ALABAMA

The committee convened four information-gathering sessions in Mobile, Alabama (Mobile County), on January 31 and February 1, 2023. The first session included representatives from a national nonprofit whose local office focuses on education, financial stability, and disaster response and recovery; the University of South Alabama; and Mobile-based CBOs with expertise in the areas of community and public health, resilient housing, emergency management, and insurance research (Alabama NGO Panel). The second session comprised city government officials from neighboring Baldwin County from the Offices of Environmental and Grants Management, Public Works, and Building (Baldwin Officials Panel). Mobile County and city officials who represented county and city environmental services, the Mobile County Health Department, the Mobile County Emergency Management Agency, the public school system, the city's Neighborhood Development Department, and resiliency efforts based in the mayor's office (Mobile Officials Panel). The fourth session included representatives from the Alabama Departments of Public Health, Insurance, and Geological Survey, and field officers from FEMA and the U.S. Department of Housing and Urban Development (HUD) (Alabama State/Federal Officials Panel). In total, 27 people participated in the committee's information-gathering sessions in Alabama.

2020–2021 Disasters of Greatest Impact

In addition to the COVID-19 pandemic, panelists agreed that the two main events with the greatest impact on their region during 2020–2021 were Hurricane Sally in September 2020 and Hurricane Zeta in October 2020. Hurricane Sally's impacts were worst in Baldwin County, while Zeta affected primarily rural communities in Mobile County. Although Sally was a lower-intensity storm, its heavy rains and sustained winds (slow forward speed) led to widespread tree falls, damage to deferred-maintenance buildings, and power outages lasting 2–3 weeks. These effects were then compounded by Zeta's arrival, interrupting and delaying recovery efforts.

Many panelists noted that the compounding effect was predominantly on the organizations and agencies serving frontline communities. Several panelists mentioned that Hurricane Ida and Winter Storm Uri in February 2021 were additional weather-related events that further interrupted and delayed recovery from Hurricanes Sally and Zeta. These counties hosted people who fled storms from neighboring regions, and many of the evacuees were coping with such problems as alcoholism, mental health issues, and severe stress, which placed additional strain on the local Alabama communities and organizations. Panelists also mentioned the ways in which previous hurricanes—Ivan in 2004 and Katrina in 2005—influenced the response and recovery processes for Hurricanes Sally and Zeta. The 15-year break between the earlier damaging hurricanes and those in 2020–2021, panelists commented, caused complacency at nearly every level, including preparation, evacuation, and response. Box 3-3 provides an example of a scenario-building workshop held in preparation for compounding disasters.

Recovery Indicators and Timing

Panelists listed short-term indicators such as debris removal (which typically is slower in more rural areas of the county), seeing fewer blue tarps on roofs, power restoration, rebuilding of homes or new construction, reengagement of social connections and networks, insurance claims being settled, schools back in session, and shelters closing. Other indicators mentioned included restoration of community functionality; hospitalization numbers returning to baseline; sediment back onshore and dunes replenished; and jobs, power, infrastructure, civic sectors, and population restored. Panelists also remarked that recovery from the events of 2020–2021 is ongoing and will take years. In addition to the public assistance projects under way or already completed, there are more in the queue, and they will take several years to complete. COVID-19 recovery indicators at the public health level included sustainable state staffing models, declining case counts, and reduced need for ventilators. Long-term health effects will be seen for years, commented one panelist, since so many people were forced to defer screenings and other medical procedures. Generally, this panelist noted, people are sicker now and have worse health outcomes.

Realized Impacts of 2020–2021 Disasters

The committee referred to the closely successive occurrence of Hurricanes Sally and Zeta, coincident with the pandemic, as a "joint pandemic event"—dubbing the events as "Zally." Because Hurricanes Sally and Zeta made landfall within a short period, some panelists struggled to separate the events and their effects when trying to recall associated specifics. Panelists

---

BOX 3-3
Scenario-Based Planning for Compounding Disasters

During the *Deepwater Horizon* disaster in 2010, the U.S. Department of the Interior launched a Strategic Science Initiative to develop a method for mobilizing real-time science guidance for decision-makers in the midst of national emergencies. The group developed a scenario-building workshop procedure that envisioned the possibility of sequential events interrupting the recovery process and thrusting communities into a situation that would require a steeper and longer climb to recovery (Machlis and McNutt, 2010). Scenarios carried out by the group considered what would happen if a hurricane blew through the Gulf of Mexico during the response to the oil spill. This type of exercise is crucial to understanding and adequately planning for the co-occurrence of multiple disasters.

In 2017, a nonprofit mental health organization in Mobile, Alabama, received a grant from the National Academies of Sciences, Engineering, and Medicine to conduct a tabletop exercise focused on a scenario in which pandemic influenza and a hurricane occurred concurrently. The exercise provided disaster-related trauma and mental health training for community members, mental health professionals, and social service providers, intended to support a train-the-trainer model that could sustain this network over time. According to a session panelist involved in the exercise, at the time, no one believed that such a scenario could occur. When the events of 2020–2021 began to unfold, however, the panelist explained that the organization was better prepared to understand the implications and address the community's needs more effectively.

---

discussed the ways in which the joint pandemic event affected the social determinants of health in Mobile and Baldwin Counties. Housing was affected in several ways. Safe, sanitary, and secure housing stock was reduced by storm damage, particularly after Sally's slow progression. One Mobile County official noted that Sally damaged 57 schools and Zeta, 34, explaining that "when you've got that kind of magnitude on the first round and then you've got to come back and determine what Sally damaged and Zeta damaged, and [deal with] insurance [and] claim adjusters," it was infeasible to disentangle which storm did what damage and where. As was the case in Calcasieu and Cameron Parishes, pandemic protocols made it impossible for insurance adjusters and FEMA staff to conduct in-person assessments; instead, they had to use photos provided by property owners to document damages. It is worth noting that opinions from the FEMA representative and some panelists vastly differed as to whether this adaptation for documenting damages ultimately helped or hindered the claims process and affected payout amounts. In rural, bayou areas of Mobile County, many houses flooded, but residents were too overwhelmed by the pandemic-era requirements to apply for assistance online.

Some landlords with homes in HUD's Housing Choice Voucher program moved tenants out to make repairs, but when they realized they could double the rent as housing stock was reduced, they disenrolled from the program, further depleting affordable housing options and forcing families to move to other areas where they could afford housing. Residents with vouchers could relocate but were reluctant to do so because of commuting distances from schools and jobs.

Coinciding with losses in the housing inventory, demand for food stamps surged, and the safety net of social services suffered. Utilities were severely compromised after what one panelist referred to as a "weird anomaly of a storm." Power outages lasted for 2–3 weeks after Sally. Many medically vulnerable, electricity-dependent residents were disproportionately affected by the power outages and reduced transportation options resulting from downed trees and closed roads. Concern was also expressed that as many as 300 jobs would be lost when federal funds, which currently pay for positions such as nurses, security, and childcare, stopped flowing into the counties.

Following hurricanes Sally and Zeta, many critical industries had difficulty finding enough workers, explained session participants. Panelists noted that many workers were resentful that they could not work from home and reassessed whether they wanted to continue to be considered "essential workers" while working in professions such as public health. There was agreement among panelists that a pathway to increasing the number of construction industry workers should include a combination of immigration policy reform and improved pay and benefits. Panelists mentioned, however, that these workforce changes were not always negative for the workers: many residents started their own home-based businesses and were able to realize entrepreneurial successes and gain more family time.

Panelists spoke about the strong sense of solidarity and willingness to help each other amid the challenges of a pandemic and other disasters. Many panelists noted that cooperation among different parishes, counties, and state and local actors was well coordinated overall for the 2020–2021 disasters. Many agreed that the disasters brought back a sense of community closeness and caring for one another that COVID-19 had taken away.

KEY THEMES FROM THE INFORMATION-GATHERING SESSIONS

Although the panelists who participated in the information-gathering sessions experienced different disasters in myriad ways in their respective areas, several recurring themes emerged from their remarks. The following sections summarize overarching messages or recurring topics voiced by panelists throughout the sessions and analyzed by the committee.

COVID-19 Complications in Disaster Preparedness, Response, and Recovery

Of course the pandemic impacted us. We have to change the way we do business, everybody did. – Galveston County Public Official

[The pandemic was a] very prolonged disaster with tentacles reaching into virtually every aspect of life, whether it's employment, changes in employment status, telework, changing how schools are being operated, kids having to do school from home. Those are all significant impacts on locations and all kinds of operations. – Harris County Public Official

A consistent theme in panelists' remarks was that the COVID-19 pandemic challenged the GOM region's historical approach to disaster preparation, response, and recovery, illuminating the emergent understanding of the concept and impact of compounding disasters. In each of the six parishes/counties, routine preparations for hurricane season were implemented and coordinated to the extent possible, yet even the best-laid plans required nearly constant adaptation and improvisation to account for the realities on the ground caused by COVID-19. In preparation for the 2020 hurricane season, state and county/parish officials worked with FEMA to determine strategies for effective and safe non-congregate shelter protocols that would accord with COVID-19 social distancing protocols. Panelists recalled that COVID-19 restrictions prevented staff from conducting annual hurricane training run-throughs that typically serve to help unify staff and familiarize them with procedures. One NGO leader asked the room, "How do you manage a hurricane during a pandemic? You're not supposed to be near each other. . . . Our volunteers went away because they were scared." Challenges encountered in the preparation phases continued into the response phase during hurricane season.

Evacuation

COVID-19 increased residents' reluctance to evacuate during the 2020 and 2021 hurricane seasons or to seek warming centers during Winter Storm Uri. Even if they were willing to evacuate, social distancing or the fear of contracting or transmitting COVID-19 prevented them from sheltering with family or friends. Fear of infection, as well as negative past evacuation experiences (e.g., restrictions on reentering Calcasieu Parish shortly after 2005's Hurricane Rita) prompted lower demand for congregate sheltering, with some households preferring to shelter in place (Collins et al., 2021). Following federal guidelines, state and local authorities pivoted from normal transport plans to reduced capacity for buses and other public vehicles used for evacuation, so that more vehicles were needed to move a fixed number of individuals. Panelists noted particular challenges in responding to medically vulnerable households in light of shortages affecting ambulance and emergency medical services. Officials had to update evacuation procedures to include the provision of personal protective equipment (PPE) to bus drivers, evacuation staff, and evacuated residents while in transit.

Non-Congregate Sheltering

Hurricane shelters used prior to the pandemic had to be reconfigured and procedures had to be updated to account for COVID-19 precautions. Evacuees arriving at shelters went through medical triage procedures to screen for COVID-19. Those who tested positive were isolated. Panelists mentioned complications regarding the functionality of shelters, including difficulty following COVID-19 social distancing and safety procedures, distinguishing COVID-19 from other illnesses, and isolating individuals who tested positive, as well as the insufficient number of volunteers and designated shelters. Increased square footage was allocated per bed space to adhere to the infection control imperative for achieving social distancing, decreasing shelter capacity in some locations by 75 percent. Even as the demand grew for shelter personnel, due to the need for more shelter locations coupled with the predictable losses of shelter staff to COVID-19 illness, the pandemic sharply reduced the numbers of volunteer personnel who were ready and willing to serve. In contrast, municipalities that customarily use schools as shelters, such as Mobile County, required minimal adaptation of standard sheltering plans because of the innate compartmentalization of these buildings such that groups and families could be safely isolated in individual classrooms.

To increase shelter capacity, and in light of the sharp decline in tourism, local authorities considered reserving blocks of hotel rooms for hurricane evacuees, yet many hotel rooms were already in use to support unhoused people because of COVID-19. Another option considered was using rooms in inpatient mental health and substance use treatment centers. The use of these options was limited, however, because switching from congregate shelters to a multitude of "hotel room shelters" was expensive. Nonprofit leaders described how they were supplementing public-sector efforts to house evacuees by providing resources for hotel room shelters, but noted that doing so became cost-prohibitive for organizations with limited budgets, especially after subsequent disasters and declining donations during the pandemic. Since capacity at many shelters was limited because of COVID-19 restrictions, the city of Houston provided shelter for unhoused families and individuals at the George R. Brown Convention Center during Winter Storm Uri. Separating people who were ill and determining who had been vaccinated (once vaccinations were available) also proved challenging. One innovation to emerge from Winter Storm Uri was to convert buses, hub houses, and nursing homes into warming centers where people could be safely physically distanced.

Panelists discussed the multiple trade-off decisions necessitated by the compounding disasters of the COVID-19 pandemic and weather-climate events. The unprecedented nature of the multiple disasters prompted one official from Lake Charles to describe 2020–2021 as "choosing the 'least worst' much of the time. There was no right way to do it. There was no handbook." In some jurisdictions, testing and vaccination sites were reconfigured for the hurricane season, at times in ways that dealt a blow to public health efforts to slow the spread of COVID-19. Preserving the functionality of area hospitals was a key priority. Attempts were made to keep PPE in reserve, but stock was limited, especially for hospitals that were already contending with inadequate supplies of masks, gowns, and hospital beds solely from the pandemic. The need for PPE for frontline disaster response personnel as well as evacuees in the shelters further strained the supply chain. Furthermore, staff had to be attentive in their personal lifestyle choices outside of work to avoid contracting COVID-19 and taking the whole team down. Vaccine mandates reduced staffing capacities as people unwilling to get vaccinated quit or attempted to change jobs, and offices were shut down because of COVID-19 outbreaks. Some employees were forced to choose between performing their emergency response–related jobs and aiding their own families as they contended with stressors such as emergency home repairs, inadequate finances, childcare concerns, and challenges related to abrupt school closures and virtual schooling. Trade-offs often resulted in loss of staff, institutional knowledge, and expertise. Inevitably, COVID-19 infected staff, forcing staff teams to miss work and quarantine; leave their positions over fears of coming to work; or at worst, pass away, further reducing workforce capacity. Professionals and volunteers alike became ill with COVID-19, and many were unable to perform their response duties. Some NGOs began to see residents who had once been donors or volunteers, now unemployed and needing support themselves. Rental assistance was provided to relieve some of the economic impacts of COVID-19, but also further burdened program management staff tasked with administering these new relief programs. These are just a few of the many trade-off decisions with which people in the GOM region had to contend during the compounding disasters of 2020–2021.

The slow and protracted recovery trajectories, impeded by periodic COVID-19 infection surges, extended and aggravated the stress experienced throughout the region, especially among those on the frontlines. FEMA's own operations in Louisiana were diminished in scope and reach. Routine procedures, such as conducting damage assessments and supporting those completing applications for individual assistance, were shifted from in-person to remote. As a result, not all Louisiana residents were able to receive FEMA funds for which they were eligible. Robust social networks that play such a critical role during recovery were less able to function in that vital capacity because of COVID-19 restrictions and social distancing. As a key example throughout the GOM region, fewer people were attending and donating to faith-based organizations (FBOs) and in turn, the FBOs were less able to fulfill their roles in recovery.

Both immediately and in the long-term, disaster resources and other forms of support are often provided by members of the community and faith- and community-based organizations. While these volunteers and organizations are critical during this time, they may be experiencing consequences of the disaster themselves and struggle to assist others. Overreliance on this often volunteer network risks the exhaustion of critical human resource capacity for effective disaster recovery.

Disproportionate Impacts on Vulnerable Populations

It's always the black and brown communities that have been historically disinvested. So, I think the data shows that we have been underserved, and we are the black and brown communities that are always the worst first, and we are the last ones that get any mitigation or anything like that. That's why we got together to try to do it ourselves. – Harris County NGO Leader

I think there needs to be some way we could help the community, especially in lower socioeconomic situations, help them prepare. – Galveston County Public Official

Vulnerable populations, including low-income households, undocumented residents, people of color, seniors, medically high-risk patients, and non-native English speakers, were often among those most affected by the compounding disasters, and a topic discussed at length by panelists in the information-gathering sessions.

Panelists talked with insight and sensitivity about subsets of the population struggling because of low-paying jobs, unemployment, housing insecurity, and poverty. They described how, for residents living "day to day" and resource constrained because of the pandemic, the impact of a late-season hurricane perpetuated a "cycle of doom." Many who sustained hurricane damage had no insurance, unreliable working hours, uncertain pay, and limited eligibility for benefits. Residents without flood or homeowners' insurance and renters, especially low-income households, faced extraordinary hurdles and lacked the resources to repair their damaged homes. Seniors, many of whom lived in homes in disrepair before the storms, saw their homes suffer even more damage. Following the 2020–2021 compounding disaster damage to housing stocks in Louisiana, state officials warn that there has never been a greater need to identify and create affordable housing for vulnerable populations (Office of Community Development, State of Louisiana, 2024).

Panelists described these community members as those most affected by disasters, with one noting, "Every disaster impacts vulnerable populations more. We keep seeing that over and over again." Another added, "It takes 10–15 years to recover from a named storm . . . and only for those who have the means to recover. We are just trying to keep the people without means from simply drowning." Another panelist on a federal/ state panel described the key findings from a research study that was under way, examining the stark economic disparities in disaster recovery based on socioeconomic status. The official laid out the study results bluntly saying, "The study spelled it right out, if you're middle class, lower-middle class, it knocks you down. If you're lower class there's no hope, you're never going to come back out of it after the second disaster. . . That's something we have to fix." As a result, one panelist described "the desperation and blank stares. People don't expect to feel better and we're seeing that a lot among our vulnerable people. Their expectations for what life should be like are so low."

Panelists in all three states visited by the committee mentioned the inequitable distribution of aid among disaster survivors and vary according to race/ethnicity and homeownership status, a point verified by research (Bowdler and Harris, 2022; Hersher and Kellman, 2021; Muñoz and Tate, 2016; Rumbach, 2023) (see also Recommendations 12 and 13 in NASEM [2023] regarding actions to increase equity in federal grant programs). Research confirms that Black or African American and Hispanic residents were approved for recovery assistance at rates of 13 percent and 28 percent, respectively, compared with White individuals (45 percent), and lower-income renters were 23 percent less likely to receive disaster recovery assistance (Hersher and Kellman, 2021).

Panelists in Texas noted how disproportionate impacts were felt at multiple levels during and after Winter Storm Uri. Regarding evacuations, customary hurricane preparations would include evacuating vulnerable populations inland, but with winter storms the size of Uri and icy roads, there was no safe zone. Given its expansiveness and impact across all socioeconomic groups, as well as its novelty, Uri presented a scenario in which even households that would normally have been "resilient" struggled. Although the storm's impacts were widespread, studies have shown that marginalized groups were still at highest risk because of their elevated sensitivity to those impacts and their reduced adaptive capacity (Hedquist et al., 2023; Nejat et al., 2022; Ritchie et al., 2022). Frontline communities were the most affected and most likely to experience sustained power losses; lack of access to food, water, and health care; broken pipes; and suffering during the event (Lee et al., 2022; Tomko et al., 2023). Power-generating plant operators initiated service interruptions to prevent collapse of the entire power grid when power demands surpassed the available supply. These blackouts occurred more frequently and for longer periods in low-income and racial/ethnic minority census tracts compared with higher-income, nonminority tracts (Lee et al., 2022). On average, 18–19 percent of the Texas population in predominantly White areas suffered a power interruption compared with 33 percent in the high-minority areas. See Figure 3-5 for the average percentage (approximate mean share) of Texas residents affected by blackouts with different ratios of minority populations.

Throughout the information-gathering sessions, panelists were asked whether subgroups of medically high-risk patients experienced care disruptions and elevated risks during the disaster events. Almost universally, panel members indicated that these individuals faced special challenges. During the peak of the COVID-19 surges in Alabama, for example, patients with cancer who were undergoing active chemotherapy or radiation treatment were not allowed to have family members or even supportive volunteers with them during their therapy sessions because of fears of contagion. In

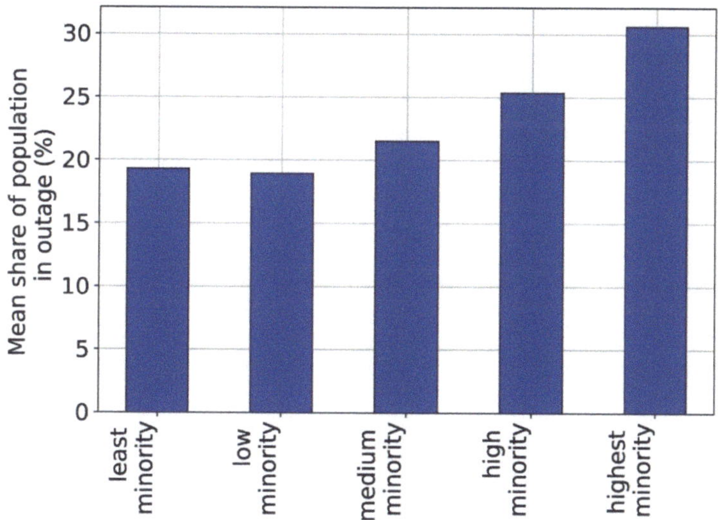

FIGURE 3-5 Average percentage (approximate mean share) of Texas population experiencing blackouts among minority quintiles during the winter storm of February 15–18, 2021. This shows that the share of Texas population experiencing blackouts increased with the share of minorities in a Census Block Group.[a]

[a] A Census Block Group is a geographical unit used by the United States Census Bureau. It is the smallest geographical unit for which the bureau publishes sample data, that is, data that is only collected from a fraction of all households (U.S. Census Bureau, 2022a). SOURCE: Shah et al., 2023.

the aftermath of hurricanes, residents who needed health care struggled to access it. Hospitals were damaged in the storms, and utility outages prevented adequate patient care.

Non-native English speakers, particularly in Bayou La Batre, Alabama, who were faced with significant flooding, water and soil contamination, and power outages, encountered equity, trust, language, and access challenges. "Disasters hit them harder," an NGO panelist explained, "and the recovery is even longer. It creates bottlenecks since their point of trust may be a single person."

The cycle of doom played out in the unhoused population, which ballooned as the limited allocation of resources was further diminished by COVID-19 and the succession of storms in 2020–2021. Many "people just moved away and did not come back," explained one NGO leader. In the aftermath, with so much unrepaired real estate, the reduced habitable housing stock led to rising rents that were unaffordable to many seeking housing at a time of economic uncertainty, underemployment, and unemployment. The departure of much of the workforce that was needed to rebuild and reconstruct houses and restore and maintain the local economy effectively served to both lengthen the recovery time and further elevate housing prices.

Panelists asserted that it is imperative to ensure from the outset that mechanisms are in place to reach disproportionately affected communities and that recovery programs integrate equitable policies and practices to address the stark disparities summarized above.

Disaster Mental Health

When Nicholas happened, some people just threw up their hands. They were so exhausted from Harvey, Uri, the pandemic. . . . It's just been one trauma after another. – Galveston County NGO Leader

One of the things that we also need to be mindful of is the trauma and the stress that was put on our responders, early responders. They were responding and responding, and they're responding again. A lot of them were responding while they were left in shambles. – Louisiana State Official

One of the major themes that emerged from the multiple panels during each of the committee's information-gathering sessions was the primacy of mental health consequences during the compounding disasters of 2020–2021, but also an appreciation for how resilience was revealed. Panelists from multiple sites provided detailed accounts of these mental health consequences, as well as evidence for regional resilience.

Panelists described one of the most egregious inequities in the realm of disaster relief with serious mental health implications—the plight of migrant populations, many of whom are undocumented and hence are disproportionately disadvantaged. Their immigration status makes these individuals ineligible for federal or state assistance, leaving them few options for accessing resources. A Harris County NGO leader stated, "There is a whole population of people whose homes can never get fully repaired, and when the next disaster strikes the grant money will be gone." Many homes with damage from disasters prior to Winter Storm Uri were further damaged by the deep freeze but were denied coverage or reimbursement from their insurance companies or FEMA after the storm struck.

In 2020–2021, many GOM residents were still grappling with protracted and incomplete recoveries from hurricanes that had occurred as long as a decade earlier. Reflecting on disasters prior to 2020–2021, panelists described many residents and community leaders as experiencing the mental health consequences of disaster fatigue. A Galveston County official, for example, stated that "response procedures were changing almost daily as COVID-19 protocols evolved." Others concurred, mentioning that COVID-19 exacerbated existing disaster fatigue and burnout. Individuals in the community experienced anxiety, trauma, fear, and PTSD due to the multiple successive disasters of 2020–2021. As noted, social networks that play a critical role in mental health and well-being during response and recovery were less able to function in that vital capacity because of COVID-19 restrictions and social distancing. A dramatic rise in domestic violence was also reported by panelists and tied to the devastating socioeconomic impacts of compounding and potentially traumatizing events. In addition, suicide rates increased among the elderly. One public health panelist hypothesized that the stress of the disasters, social isolation, and lack of resources potentially contributed to the deaths of many senior citizens, especially those on fixed incomes. The prolonged and worsening pandemic also took its toll. By the time the wave of the deadly Delta variant arrived in early October 2021, one panelist recalled that it was "as much a mental health impact as a physical impact. . . . People were exhausted and tired after being on response months and months earlier. . . . So, you kind of throw your hands up in the air . . . you're almost robotic at that point." **Comments from panelists revealed that mental health issues and chronic and acute**

stress stemming from exposure to compounding disasters are not being adequately addressed.

Panelists also discussed the mental health of students in the storm-ravaged counties. Students' education was disrupted during the pandemic, with classes shifting to online, remote delivery. COVID-19–induced educational challenges were sometimes punctuated by school closures brought about by structural damage and power outages following hurricanes and other severe weather-climate events. Many youths experienced a combination of negative emotional and educational consequences. In Louisiana, youth mental health surfaced as a major theme, in part because of the irony that students returned to in-person school only 1 day before Hurricane Laura hit and caused extensive damage to every educational facility. The storm damaged internet connectivity, severely limiting students' ability to attend virtual classes. Prevalent anxiety symptoms and pervasive social isolation characterized the students' experiences. Meanwhile, for parents with frontline response roles, an additional complication was balancing their work demands with managing their children's roller-coaster educational progress. The ripple effects of multiple disaster impacts throughout the GOM region, exacerbated by COVID-19, depleted the resources and personnel needed for recovery for all affected communities. In turn, the slow and protracted recovery efforts, hobbled by periodic COVID-19 stress, extended and aggravated the stress experienced throughout the region, especially for those charged with response duties. This piling on of pandemic and disaster stressors took a toll. One panel participant stated, "It paralyzed some [workers] as these issues compounded." Panelists expanded upon how the mental health of the frontline responders was progressively "exhausted," leaving many of them "burnt out and struggling emotionally." It was difficult for many frontline workers to ask for help. Many did not know whether—and what—resources might be available to provide support. A panelist highlighted the extreme stress and strain experienced by first responders whose work demanded unrelenting exposures to potentially traumatizing events. All types of frontline personnel were overwhelmed. One panel member commented, "From a health department perspective, this period of events will break anybody."

When prompted, some panelists were able to identify positive outcomes and evidence for community resilience, citing a strong sense of cohesion and readiness to assist one another. Panelists agreed that the concerted response to the weather disasters reinvigorated a sense of community closeness and care for one another that COVID-19 had taken away.

Taken together, the information-gathering discussions in all three states revealed the primacy of mental health consequences for pandemic and storm survivors, the heightened mental health risks for frontline responders, the disproportionate risks for marginalized communities, and the amplification of mental health risks for populations confronted by compounding disaster events occurring simultaneously or in quick succession.

SUMMARY OF KEY FINDINGS

The weather-climate events and the COVID-19 pandemic exceeded historical benchmarks and fundamentally challenged conventional approaches to disaster preparation, response, and recovery. Throughout 2020–2021, the emergence and spread of COVID-19 transformed the public health risk of compounding disasters in the GOM region by amplifying health-compromising exposures and underlying vulnerabilities while modifying emergency procedures for weather-climate events. With the onset and aftermath of each successive disruptive event, the effects within communities compounded and recovery processes became prolonged. As a result, community resilience and adaptive capacity declined, compassion fatigue set in, and key leaders struggled against complacency in their inability to deliver effective recovery assistance. Socially vulnerable populations, particularly communities of color, those of low socioeconomic status, and non-native English speakers, bore the brunt of compounding disasters. Generally, GOM communities were not prepared to respond or recover from the occurrence of disasters that overlapped in both time and space. The impacts and respective recovery time lines of each event made it difficult, if not impossible, for affected community members to differentiate between or ascribe specific impacts to each disaster. Beyond the immediate damage and loss, chronic and acute stress resulted in widespread mental health challenges among community members and responders. These effects are not adequately understood or addressed, potentially resulting in the diminishment of adaptive capacity.

4

Interdependent Systems in the Context of Compounding Disasters

Compounding disasters have widespread adverse and unintended effects on both communities and the interconnected human–natural systems on which they rely to function. Compounding disasters "co-occur such that they concurrently affect interdependent critical infrastructure systems, thereby presenting multiplicative risks to the interconnected systems and population" (Wells et al., 2022, p. 2).[1] When a community experiences a disaster amid ongoing disaster recovery processes, there can be a "greater burden on individual and interconnected networks when compared to a singular threat occurring in isolation" (Wells et al., 2022, p. 2).

A SYSTEMS APPROACH TO UNDERSTANDING COMPOUNDING DISASTERS

There are multiple approaches to examining disasters and their compounding impacts. For this chapter, the Committee on Compounding Disasters in Gulf Coast Communities, 2020–2021, adopted a systems approach to account for the contemporary drivers and realized impacts of risk (UNDRR, 2022); the complex and interconnected dynamics that underpin society; interdependencies of critical physical, virtual, and social

---

[1] In this chapter, the term *system* can be likened to a "sector" of society critical to its functioning, and the two terms are used interchangeably throughout. The former term does not imply mechanistic social decisions or actions.

149

infrastructures; and the disaster-induced amplification of underlying vulnerabilities.

As climate change heightens the potential for extreme weather-climate events to occur contemporaneously (NASEM, 2016), a holistic approach is needed to unpack the interdependencies among various services (and service providers) vital to community functioning. Using a systems approach to synthesize primary and secondary information sources—the committee's information-gathering sessions, and the commissioned papers and relevant literature—this chapter focuses on ways in which compounding disasters in the Gulf of Mexico (GOM) region during 2020–2021 affected critical social, physical, and virtual systems whose networks, assets, and services are interconnected and nested within larger systems.

The systems that serve both the routine and the acute needs of a community provide services through core functions and elements that may rely on other systems to function. Thus, service delivery depends not only on the functionality and capacity of these elements but also on the built infrastructure and social assets on which they rely. These assets may be natural (e.g., waterways, protective dune systems), constructed (e.g., bridges, electrical grids, schools), or social (e.g., institutional or lived experience, knowledge, social networks and cohesion).[2] These systems have one or more connections that (1) link their various components; (2) interconnect different services; and/or (3) connect them to larger regional, national, and even international networks. These connections include information channels (mechanisms for sharing data and directives, as well as social networks and media), supply chains, and services.

This chapter examines interdependencies among systems, service providers, and services vital to community functions, including disaster response and recovery. **These sectors must collaborate to deliver vital services in support of a community's success in planning for, responding to, recovering from, and mitigating the impacts of compounding disasters.** Their capacities and underlying infrastructure have been honed over generations of disaster experience in response to the most common hazards in the GOM—hurricanes and floods. That historical experience shapes the policies, practices, and regulations that guide how GOM communities are built, how they prepare for and respond to these threats, how

---

[2] Social networks reflect social capital and build community cohesion and do not operate as mechanistic systems. They are included here to acknowledge their function as social assets.

they direct their limited resources to build social and infrastructure resilience, and how some populations see greater disaster recovery assistance and have shorter recovery times than others.

Highlighting the key dynamics of these interacting systems and services will help shed light on some of the experiences of affected populations during the compounding disasters of 2020–2021. Analysis of cross-cutting themes from the committee's information-gathering sessions in Texas, Louisiana, and Alabama (see Chapter 3) illustrates the connections and interdependencies among these systems and services in the context of compounding disasters, including the COVID-19 pandemic, as well as the characteristics of the vicious and virtuous cycles[3] that form within and across systems. Vicious cycles are one way to show how the effects of disasters can compound.

THE PANDEMIC, VULNERABILITIES, AND SYSTEM INTERDEPENDENCIES

The duration and scope of the COVID-19 pandemic, a hazard that combined a biological threat with underlying vulnerabilities, extended the time line over which disasters could compound (Tozier de la Poterie et al., 2022) **and created unique constraints by encouraging isolation, which impeded disaster risk management actions that traditionally require a collective approach.** This section examines these tensions and the system-level effects of the compounding disasters across the GOM region, as reported by participants in the information-gathering sessions. Notably, those narratives offered limited commentary on the effects of the observed climate- and weather-related events on COVID-19 infection rates and associated health outcomes during the recovery period, likely for two reasons. First, the methodology adopted for the information-gathering sessions was event based, focusing participants on specific weather-climate events. The COVID-19 pandemic was framed as the backdrop against which these events occurred, and the sessions were structured to reference specific hurricanes, winter storms, and floods chronologically, triggering a focus on the details of the impacts of and recovery from these events. As one Alabama nongovernmental organization (NGO) leader noted when

---

[3] Vicious and virtuous cycles refer to complex chains of events that reinforce feedback loops, in which a change builds up over time. A vicious cycle is a negative reinforcing feedback loop, while a virtuous cycle is a positive reinforcing feedback loop.

explaining their hyperfocus on these events, despite the ongoing pandemic, "It's like you have a chronic illness and you have a cold. I deal with the cold, but I still have a chronic illness."

Second, because of the acute nature of each weather-climate event, session participants admitted that their attention was focused on meeting immediate human needs for water, food, and shelter, as well as socioemotional support. In addition to physical disruption of testing centers and reduced public health capacity in the aftermath of these events, participants focused on meeting those basic needs admitted to deprioritizing social distancing, testing, and even acting on the symptoms of COVID-19. Their experiences demonstrated how the pandemic, which previously had demanded their vigilance, was neglected or deprioritized when hurricanes hit, even as cases were surging, given the urgency of other basic needs such as food, water, and shelter. In the words of one Louisiana respondent, "COVID did not exist in Cameron Parish [in] August, September, October, [and] November, because people were too busy and too poor to worry about COVID." These realities could not only explain the limited emphasis on COVID-19 in the information-gathering sessions but also may suggest that COVID-19 cases were underreported during the early stages of recovery despite the potential for hurricane-related surges (Naqvi et al., 2023).

Local Economy, Civil Society, and Public Services

The pandemic spurred a nationwide economic downturn before the Atlantic hurricane season began on June 1, 2020, with the U.S. gross domestic product falling 8.9 percent in the second quarter of that year— the largest single-quarter contraction in more than 70 years (White House Council of Economic Advisers, 2022). Session panelists remarked on the economic downturn's multifaceted effects on their communities' ability to plan for, respond to, recover from, and mitigate the impacts of the sequence of disruptive events they faced. **The pandemic led to unemployment, resulting in job losses and difficulty paying bills in advance of the 2020 hurricane season, reducing certain households' adaptive capacity, and increasing sensitivity to the approaching storms.** One panelist working for an NGO in Baldwin County, Alabama, described it thus: "Many work-ing families that always were able to provide are now not able to provide. And they don't know how to ask for help because they've never had to ask for help. And our resources were slim . . . it was every area of life, every work field . . . it was everything. The calls were nonstop, 24 hours a day."

Panelists noted the vital role of community-based organizations (CBOs) and faith-based organizations (FBOs) in delivering aid. CBOs and FBOs across the GOM region play an especially prominent role. From local churches to national groups with emergency response teams, these organizations are on the ground quickly and attend to local needs—from food to basic household items to rebuilding of houses to counseling—with compassion. Even so, church membership has been declining in the region (as nationally), particularly during the pandemic. This decline translates into fewer donations and fewer volunteers to assist with aid programs. **In the future, CBOs and FBOs may no longer be able to provide vital services as they have in the past as sequential and compounding events strain volunteer staffing even more than singular events.**

The economic and social instabilities' effect on public and civil society services was twofold: the economic downturn tempered donations to nonprofits, FBOs, and other entities in the civil society sector and reduced tax revenues for local governments, which in turn reduced overall revenue streams. At the same time, community members' needs for food, mental health counseling, and emergency rental and utility assistance were expanding. Those in need increasingly looked to CBOs and FBOs as trusted sources of information and resources and key pillars of disaster risk management. Their calls for assistance increased at a time when many CBOs and FBOs were forced to reduce the scale of their operations or close because of pandemic restrictions. In some cases, charitable services were disrupted. Many of those that remained open struggled to adapt their operations to virtual or hybrid mode or limited in-person staffing. Essential services were forced to remain operational, often in staggered shifts, but one COVID-19 infection among staff could shutter offices critical to recovery, such as those responsible for building permits. As a result, the capacity to deliver critical services in response to rising demand inevitably suffered. Mounting demands due to the subsequent sequence of disruptive events in each community only accelerated burnout among limited staff members, who were often also contending with the same shocks in their personal lives. Staff turnover ensued, further reducing efficiency, as onboarding taxed these providers' already limited capacity.

Organizations struggled to absorb the staff shortages, in part prompted by the pandemic's "great resignation"[4] (see Table 4-1 for 2019–2021

---

[4] The "great resignation" refers to the trend of large numbers of employees voluntarily quitting their jobs. The term was initially coined by Texas A&M professor Anthony Klotz to describe the mass resignation that occurred at the beginning of the COVID-19 pandemic (Phipps, 2022).

TABLE 4-1 Unemployment in Selected Counties/Parishes, 2019–2021

Unemployment in Selected Counties/Parishes, 2019–2021			
County/Parish	% of Population Unemployed, 2019[a]	% of Population Unemployed, 2020[b]	% of Population Unemployed, 2021[c]
Galveston County, TX	4.1	8.8	6.6
Harris County, TX	3.9	9.0	6.4
Texas Average	3.5	7.7	5.6
Calcasieu Parish, LA	3.9	9.5	5.7
Cameron Parish, LA	3.6	6.1	3.6
Louisiana Average	4.6	8.6	5.6
Jefferson Davis County, MS	7.0	9.5	7.5
Marion County, MS	5.6	7.0	5.2
Mississippi Average	5.5	8.0	5.5
Baldwin County, AL	2.9	6.1	2.9
Mobile County, AL	3.9	8.6	4.6
Alabama Average	3.2	6.4	3.4
United States Average[d]	3.7	8.1	5.4

NOTE: Percentages are annual unemployment rate averages.
[a] U.S. Bureau of Labor Statistics, 2023a.
[b] U.S. Bureau of Labor Statistics, 2023b.
[c] U.S. Bureau of Labor Statistics, 2023c.
[d] U.S. Bureau of Labor Statistics, 2023d.

unemployment statistics). Voluntary withdrawal from the workforce due to burnout, illness, better/safer opportunities, or the demands of homeschooling/childcare/caregiving drew staff away from critical sectors such as health care, education, and public service, as well as debris removal and construction—all essential for disaster preparation, response, and recovery. Workforce shortages in turn drove up competitive wages and the demand for benefits in many sectors, which, while advantageous for workers, further strained the limited revenue available to public and civil society service providers.

In addition to the paid workforce, unpaid volunteers represent a substantial portion of the workforce in both the civil society and public-service sectors. They staff human and animal shelters, vaccine clinics, and polling places, as well as volunteer fire departments, vital to the pandemic response in such areas as rural Mississippi (WJTV, 2020). Volunteers are often older retirees, whose age puts them at higher risk of death or severe illness from COVID-19. Thus, the depletion of the volunteer workforce, many of whom were elderly, further reduced the capacity of public and civil society service providers. Many panelists reported that the volunteer force had not yet fully returned.

The global economic downturn and workforce shortages also spurred supply chain shortages that, beyond an inability to access key commodities such as medical supplies, impaired the ability to deliver services effectively. Pandemic-induced strains on supply chains were exacerbated by subsequent disasters, inflation, and rising regional demands for critical commodities during relief efforts (e.g., prices on food and home rebuilding and repair skyrocketed). Residents were then faced with going "to a local hardware store to buy a sheet of sheetrock, and it was twice the price in Lake Charles than in Moab or Houston, which made economic recovery very challenging because prices were so inflated." Another session panelist in Lake Charles explained that "we were not just dealing with trying to recover from the two storms back-to-back, which is a battle of its own. COVID-19 fundamentally altered the world economy, and all that was happening at the same time." Whether because of the added cost of operations due to physical distancing, testing, quarantining, and sanitation protocols and/or the overall rising costs of other necessary operational commodities, public and civic society service providers were forced to further scale back the volume of services that could be delivered under revenue streams already reduced by the pandemic and further impaired by subsequent disruptive events. Panelists noted that the effects were particularly pronounced for building materials after the sequence of hurricanes in 2020–2021.

Housing

Housing plays a central role in disaster risk management and public safety, as well as overall well-being. At the same time, disasters have a long-term negative effect on housing perhaps more than on any other aspect of the built environment. **The disasters of 2020–2021 underscored how housing collectively functions as vital infrastructure to sustain education, social cohesion, and economic activity in times where the physical facilities originally designed for these functions cannot be occupied.** Thus, the unaddressed vulnerabilities of housing influence the sensitivity of a community or individual to compounding disasters. Substantive impacts to the housing sector also cascade into other sectors vital to recovery and societal function.

Impacts of Inadequate Disaster Recovery Processes

Inadequate and event-based disaster recovery processes at the federal, state, and municipal levels magnify the impacts on housing in compounding disasters. Because of pandemic restrictions, federal disaster recovery assistance requests and insurance claims to repair damaged homes were processed virtually. Inefficiencies abounded, and residents had to navigate these new systems without the customary on-site, in-person support. Mail service challenges delayed the distribution of assistance and benefits, in turn delaying debris removal and restoration of critical municipal infrastructure while halting housing repairs and reconstruction. Individual households and municipalities experienced a similar vicious cycle driven by the sudden adaptation of existing systems for managing assistance requests and insurance claims during a pandemic (see ① in Figure 4-1). Moreover, those municipalities and households that most need recovery funds are often those least well equipped to navigate complex and continually evolving online recovery assistance systems.

Session participants in Alabama and Louisiana criticized the remote interactions that replaced the customary face-to-face encounters of disaster recovery, which added to the stress, anxiety, and confusion among storm survivors. This was particularly evident among the elderly, who generally were less proficient in some of the newer technologies and were unable to complete forms to request postdisaster assistance. Residents experiencing homelessness similarly faced substantial challenges in obtaining access to technologies. In short, unable to directly interact with trusted local partners who are critical in helping navigate the bureaucracy of the federal system (Trivedi, 2020), the most vulnerable populations were particularly affected in the virtualized world of the pandemic.

During a time of protracted recovery, affected households and their communities are increasingly vulnerable to future disruptive events, creating the conditions necessary for a second vicious cycle to form (see ② in Figure 4-1). Disaster recovery systems that traditionally associated losses with discrete events were unable to process losses that were now due to the cumulative damages incurred during sequential hazard events. This led to delayed recovery processes, leaving damaged buildings and infrastructure unrepaired and eventually further damaged in subsequent events, creating

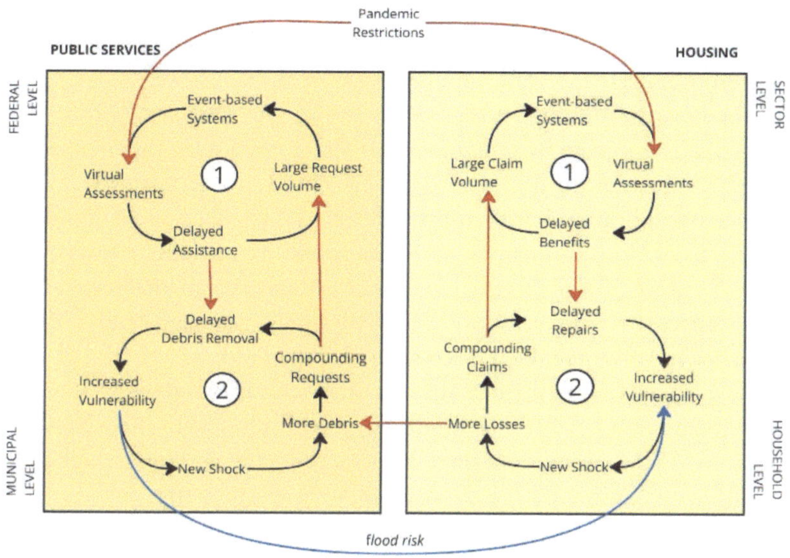

FIGURE 4-1 Systems map of the interdependencies between housing and public services in recovery.

a vicious reinforcing loop between household and community recovery processes.

Homeowners confronting these compounding cycles of housing losses also had to bear the accumulated burden of repeated attempts to engage with Federal Emergency Management Agency (FEMA) officials, insurance adjusters, plumbers, electricians, roofers, and contractors for home and building repairs. Panelists noted that many of the individuals working in these capacities were also helping to repair their own homes or those of neighbors or loved ones and felt the strain of the higher demand for their services in this 2-year period. For homeowners with more than one insurance policy, attempts to recoup losses could be multiplied by the number and type of insurance they carried, from flood to windstorm to homeowners' insurance for their homes, in addition to insurance for vehicles, boats, and businesses that were also affected by these events. With deductibles enforced by each policy for each event, many households quickly exhausted whatever financial reserves were available to them. As Underhill (2009, p. 62) notes, such "mounting disaster expenses may exacerbate people's already vulnerable economic situation, making it difficult for some to keep up with pre-existing bills and newly emergent expenses." As one panelist noted, "Being resilient also means . . . folks just keep taking the hits over and over again . . . it becomes exhausting after a while."

The virtualization of damage assessments by insurers and FEMA often required self-reporting by individuals with no formal training or expertise, leading to omission of necessary data that could not be retrieved if clarifications or additional information were requested. One workaround was FEMA's use of drone imagery and windshield assessments for damaged homes. However, the inability to conduct up-close assessments led to errors. For instance, the overspecification of substantially damaged properties in Lake Charles left repairable houses slated for demolition.

Beyond the inconvenience of a complete rebuild was the loss of grandfathered base flood elevations, forcing new construction to meet elevation requirements that were more than 50 percent higher, at considerable cost. In Southwest Louisiana, the reliance on these perfunctory posthurricane inspections instead of in-person inspections received harsh criticism. A Lake Charles city official explained that, eventually, city officials organized their own reassessment effort using local engineers to reverse the windshield assessments, though this further extended the time damaged homes were vulnerable to subsequent shocks.

The sequence of events as described by session participants working and residing in the Lake Charles region demonstrates some of the inadequacies of the nation's current disaster recovery process, which can result in prolonged exposure in a state of increased vulnerability. In some cases, the property assessments were not completed before a subsequent disruptive event occurred. The current event-based disaster recovery system assumes that losses are associated with discrete events. But when damages from one event have not yet been assessed or repaired before the next event creates new, compounding losses, complications ensue.

Panelists from the Lake Charles region described the ways in which such event-based disaster recovery processes compounded the 2020 disaster impacts. Uncleared debris from Hurricane Laura was mobilized by Hurricane Delta's wind and water, worsening the damaging effects of both storms. Homes with damaged roofs from Hurricane Laura were reexposed as their tarps were blown off by Hurricane Delta. Interior water damage from this second storm, enabled by the unaddressed wind damages from Hurricane Laura, created more complex claims scenarios that event-based reporting systems were ill equipped to handle. The damages from the now-compounding disasters added to the volume of debris to be managed, both physically and with respect to assistance requests, further delaying the flow of resources to municipalities and households alike (see Figure 4-1 for an illustration of this vicious cycle). As one panelist noted, "That's one of the problems with the way the [FEMA] system is designed . . . there are these . . . mass processes . . . [and] they expect every disaster to fit . . . in this mold." Municipalities struggled to navigate FEMA's event-based reporting structures and its implicit expectation that debris from sequential events could somehow be distinguished.

In the May 2021 floods in the Lake Charles region, debris not yet removed from the 2020 storm sequence clogged stormwater systems in the affected areas, limiting the capacity available to manage the record rainfall. Areas flooded that had never flooded before, bringing losses to families not required, or who could not afford, to carry flood insurance. Hurricane-damaged houses awaiting repair were tarped with materials not rated for drastic changes in temperature or long-term use, making these water barriers porous and vulnerable to ongoing water penetration when the heavy rains arrived. The floods generated even more debris and added yet another round of losses and claims to an already overwhelmed claims and assistance system that was unable to quickly mobilize funds into Lake Charles.

The arrival of Winter Storm Uri in the GOM region presented a challenge that affected both unoccupied homes and houses with unrepaired damage from previous disasters. Some panelists highlighted that slow-to-recover properties in Texas, still stripped to their studs after Harvey (in 2017), met Uri without insulation. A similar risk was posed in Galveston County, which, like other tourist destinations on the Gulf Coast, has a high number of unoccupied rental properties in the winter months. Whether these properties were stripped bare or unoccupied, the end result was the same: they were not heated during the deep freeze; pipes ruptured tapping municipal water supplies, even more water damage was incurred by yet-unrepaired homes, and an inventory of rental properties vital to residents and the local economy was damaged.

Despite a general "all hazards" approach in the United States, Gissing and colleagues (2022) report that traditional risk management almost always focuses on discrete events. Compounding disasters present challenges of greater scale and complexity relative to individual events, and the repository of lessons learned is just being studied (see Chapter 5). Failure to adequately consider the effects of compounding events can slow recovery, increase reconstruction costs, and accentuate economic and psychological stress (Gissing et al., 2022). While all-hazards planning enables preparations for either natural or technological hazards, it usually misses the compounding impacts and the impacts that linger during the long recovery period.

Impacts on Affordable Housing

In the housing sector, as in so many aspects of disaster planning, preparation, and recovery, marginalized populations bear the brunt of disaster impacts. This is particularly evident in the effects of these events on the supply of affordable housing.

Louisiana's Calcasieu and Cameron Parishes saw significant damage and loss of housing supply following the 2020 hurricane season. Hurricanes Laura and Delta damaged approximately 50,000 housing units in Calcasieu Parish (McKinsey and Company, 2020). As the supply of habitable homes was reduced, rental prices increased, and the unhoused population ballooned. Meanwhile, the overall socioeconomic decline in Alabama's Baldwin and Mobile Counties, well before 2020, had resulted in a corresponding decline in the condition of the physical housing inventory, leading to more deferred-maintenance structures, as well as shortages of affordable housing. These were early indicators of not only the physical vulnerabilities of the housing inventory but also the dysfunction of the housing market system across all three states visited by the committee.

Many hurricane-damaged rental properties were deprioritized for repair, absorbed by buyout programs, or flipped and sold by their landlords, while others were deemed unrepairable by the homeowner because of insufficient insurance coverage and available savings. Tenants in rental properties subsidized by the Department of Housing and Urban Development (HUD) could be displaced for repairs, but many property owners soon realized a restored house's value in the inflated housing market, withdrawing it from HUD-supported programs to permanently displace the tenants. All these pathways resulted in the same outcome: the stock of affordable housing dropped after each weather-climate event, often permanently displacing low-income households in most need of housing (Brennan et al., 2021; see Figure 4-2). Affordable and/or government-subsidized rentals are slow to return (if ever) to the market after a weather-climate event. This has wide-ranging and long-standing impacts, especially if more than one event occurs adjacent in time and/or region (Rumbach and Makarewicz, 2017; Van Zandt and Sloan, 2017).

These system dynamics had wide-ranging consequences, as much of the critical workforce needed to rebuild, restore, and maintain the local

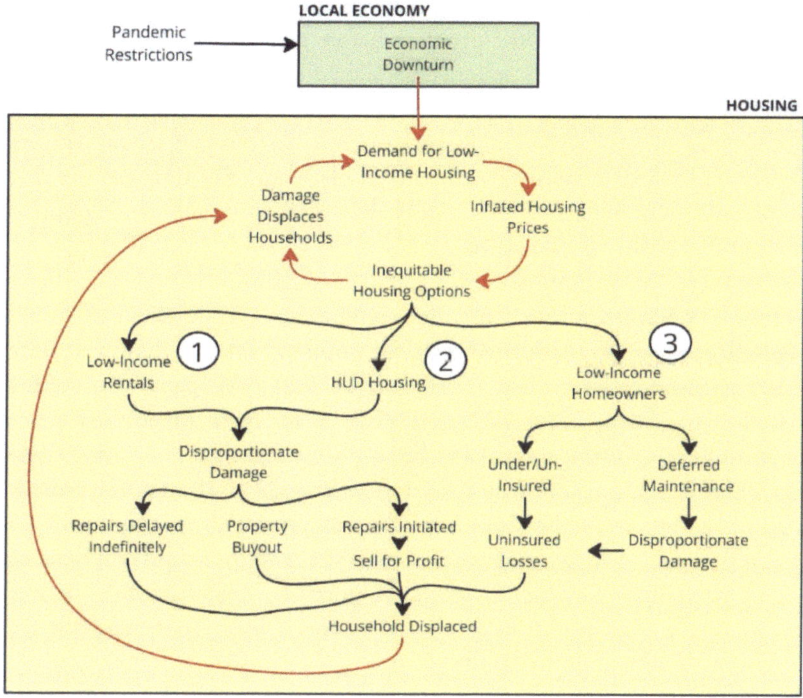

FIGURE 4-2 Systems map of vicious cycles spurring an even higher demand for affordable housing and more displaced households.

economy could not afford to rent, much less purchase, a home in the area. These disruptions further depleted the workforce vital to the recovery of the local economy. As one session participant stated, "Housing and [specifically] affordable housing is what is going to keep that workforce away." Moreover, since mental and physical health can be compromised by unsafe living spaces, loss of housing stock can also increase impacts on the health care sector. Calcasieu Parish's program director for disaster housing recovery and chair of the Lake Charles Housing Authority Board of Commissioners noted further that "we are very slow to come back with units that are truly affordable. . . . Even a schoolteacher is going to have a hard time finding a unit that's affordable" (Maschke, 2023).

Participants in the panel on housing and displacement emphasized that these dynamics are well known, and even if communities lack the funds to address the affordable housing crisis proactively, they can maximize the effectiveness of their postdisaster resources by planning their housing recovery programs to deal with the problem before a disaster occurs. Awareness of the highest-risk areas/households can allow the identification of locations where temporary and, ideally, permanent affordable housing should be mobilized after a disaster to maintain proximity to jobs and schools. Panelists emphasized that having agreements in place ahead of time ensures that communities can take swift action to stabilize the most vulnerable populations once recovery funds arrive.

Panelists from Southwest Louisiana mentioned that state agencies and FEMA have continued to offer support, and some panelists found themselves pleasantly surprised that the support did not disappear with Hurricane Ida's arrival in that area in August 2021. One panelist suggested that the amount of disaster housing assistance offered for lower-income neighborhoods has the potential to be "transformational" over time, sparking hopes that the recent infusion of disaster recovery funds will prompt not only a long-awaited reset on the affordable housing crisis in the GOM region, but a reset that strengthens the resilience of the affordable housing inventory to future hurricanes, floods, and tornadoes. (See Box 4-1 for more information on flexible funding related to affordable housing and Table 4-2 for county/parish appropriations.)

One panelist from Mobile County similarly remarked that "from an affordable housing perspective there is a shift that is coming; it has definitely started in Louisiana, where the HUD money that has been coming in post-Laura and [post-]Ida is truly building affordable housing that can take a hit and still be secure housing, and it is slowly moving its way [to Alabama]." Panelists noted further that one way to expedite this process is to shift from compensating the market value of a physically vulnerable structure to financing the cost of its replacement with an improved property, such as one with an IBHS (Insurance Institute for Business and Home Safety) FORTIFIED roof, hurricane-grade windows, and elevated foundations above the requisite base flood elevation, or incentivizing relocation outside of the Special Flood Hazard Area.

Health Care and Education

New modes of virtual service delivery, including major developments such as telehealth, virtual counseling, and e-learning, became staples of routine service delivery during the pandemic and proved vital as the later sequence of weather-climate events further disrupted in-person services.

BOX 4-1
Federal COVID-19 Relief Funding:
An Unexpected Recovery Accelerator

Several session panelists noted that it would be difficult to project what their community's recovery experiences might have been like without the infusion of funds authorized by the American Rescue Plan Act (ARPA; P.L. 117-2). Most session panelists referred to federal COVID-19 Relief Funding simply as "COVID money," without distinguishing between funding streams. The Coronavirus State and Local Fiscal Recovery Funds, or SLFRF, program provided funding to state, territorial, local, and tribal governments across the country to support their response to and recovery from the COVID-19 public health emergency.

With disaster recovery funds delayed, the flexibility and "pre-positioning" of ARPA funds led some (not all, as several panelists noted) local governments to unexpectedly fill acute gaps in the housing sector through emergency rental and utility assistance. This flexibility prompted panelists to note a renewed focus on adapting policies quickly and critically, looking at ways in which federal and other funds are utilized by local and state governments. ARPA funds were eventually used to harden infrastructure and improve flood management capacity, addressing preexisting infrastructure deficits in areas that contributed to disaster risk reduction. Panelists in both Mobile County and Calcasieu Parish noted the critical importance of COVID-19 relief funding in providing rental and utility assistance to housing-insecure residents. However, in many cases, these funds were a stopgap that masked unmet needs for vulnerable populations following the compounding disasters.

Yet, while these digital modes had broad reach in most states outside the GOM region during the pandemic, panelists noted that residential broadband access of their counties and parishes was below national averages, limiting households' ability to support e-learning and telework during the extended closures resulting from the pandemic and damaging weather-climate events. Many schoolchildren lacked or had inadequate internet access; those who did have access often faced other challenges, such as an insufficient number of computers given the number of remote workers and learners in their homes. The end result was the same:

TABLE 4-2 Total Appropriations from the ARPA, by County, as of May 2021

Funds Authorized by the ARPA, Appropriations by County/Parish	
County/Parish	**Allocation ($)**
Galveston County, TX	66,456,490.00
Harris County, TX	915,508,128.00
Calcasieu Parish, LA	39,515,058.00
Cameron Parish, LA	1,354,424.00
Jefferson Davis County, MS	2,161,484.00
Marion County, MS	4,773,017.00
Baldwin County, AL	43,360,588.00
Mobile County, AL	$80,261,198.00

SOURCE: U.S. Treasury Department, 2021.

children struggled to follow their lessons remotely. The loss of power and the resulting disruption to broadband services that occurred during the winter storms or subsequent hurricanes ultimately exposed the growing social dependence on these critical communication technologies and their fragility.

IMPACT OF REGIONALITY ON DISASTER RECOVERY

The GOM region has a high level of interconnectedness. While all regions have some degree of interdependence due to their reliance on common supply chains or federal services, reciprocal agreements and evacuation practices across the GOM region draw on the shared capacity of state and local governments and NGOs to help communities in neighboring regions absorb and recover from large-scale disasters, particularly hurricanes. **Resource sharing can provide a robust response to a single geographically focused event such as a hurricane. But when major events occur in succession, which happened during 2020's hurricane season, this interconnectedness can instead cascade the adverse effects across the region and stall ongoing recovery processes.** Diversion of recovery resources to other affected areas reinforces the "regionalism" of disaster effects.

Regionality and the Pandemic

The COVID-19 pandemic was global in scope and, unlike a hurricane, had wider regional impacts that made drawing on resources from neighboring jurisdictions impossible in some cases, especially for essential supplies and personal protective equipment (PPE). Shortages of such supplies were nationwide, as the pandemic impeded international and national supply chains for many resources. Sanitation and building supplies, medical equipment, and other tools vital to recovery were in short supply, and localities either had to wait to get to the head of the queue or bid competitively with other localities for the same supplies.

Beyond supply chains, the pandemic compounded the inability of regional assistance networks to function as normal. It depleted the number of emergency and medical personnel available to respond outside their normal bases of operation, and the loss of those individuals facing quarantine because of COVID-19 infection and those unable to leave an already strained situation at home further reduced the capacity for regional cooperation.

Supply Chains and Essential Supplies and Workers

The disasters and recovery sequences described by session panelists emphasized how physical damage resulted in disruption of service delivery. However, panelists in each session emphasized the regional effects of hurricanes on neighboring communities. These effects included the need to host and support evacuees, which can tax already strained services and sectors. According to one NGO panelist, "When they evacuate here . . . we're trying to find housing for them, we're trying to help them with all their myriad of life issues that they bring with them when they flee the storm. And it puts an already stressed nonprofit community into further stress." Given that populations who move across county/parish and even state lines may never return, their migration has lasting (and potentially adverse) ramifications on the systems serving both their new and former communities.

Governments (federal, state, and local) and nonprofits have reciprocal agreements or mandates to support responses across the GOM region, drawing personnel and resources away from ongoing recovery in one locality to support acute situations in others. The sequence of storms in the region prompted shifts in priority toward places suffering from the most-recent events (e.g., from Lake Charles after Hurricane Laura/Delta to Houma after Hurricane Ida). These shifts prompted both perceived and real shortages in the places with less-recent damage. Other critical commodities and services related to reconstruction, including the supply of line workers and construction materials, also had to be shared across the region, with spikes in demand raising prices to further delay the ongoing recovery process.

While other GOM counties and parishes were in various stages of disaster response and recovery, Mobile and Baldwin Counties saw their building supplies reduced "for probably the next 3 years just because of Cat 4 Ida in LA," explained one participant. Relatedly, when insurance companies in neighboring Louisiana began collapsing, pulling out of the state market, canceling policies, and generally causing market chaos, Alabama started to see similar effects, explained panelists.

Winter Storm Uri also made seeking aid from neighboring communities nearly impossible as subfreezing temperatures made roads impassable, with many areas encountering ice, and the advance shutdown of many public transportation options (Castellanos et al., 2023). Winter Storm Uri thus illustrated what hurricanes historically had not: a single event whose geographic reach was sufficient to impair this shared capacity across the region, delaying recovery processes at all levels. Beyond building supplies, the committee heard how strain brought about by ongoing regional disasters negatively affected disparate county resources such as staffing for NGOs and other organizations; donations; or the availability of volunteers, PPE, power restoration, construction workers, and more.

SUMMARY OF KEY FINDINGS

The duration and scope of the COVID-19 pandemic, a hazard that combined a biological threat with underlying vulnerabilities, extended the timeline over which disruptive events and disasters could compound. Disasters occurring in 2020–2021 inhibited the ability of regional supply chains and federal, state, and local support networks to function as normal. Generally, cooperative resource sharing can provide a robust response to a single geographically focused event such as a hurricane. However, when disruptive events occur in succession, which happened in 2020–2021, this interconnectedness can instead cascade the adverse effects across the region and further stall ongoing recovery processes.

The pandemic led to extensive unemployment and economic hardship in advance of the 2020 hurricane season, increasing the vulnerability and sensitivity of communities throughout the region to the effects of the approaching storms. Storms caused widespread damage throughout the region, resulting in significant uninsured losses and supply chain disruptions that significantly increased the cost of rebuilding materials. As economic hardship increased and persisted, communities began to destabilize in the aftermath. Faith- and community-based volunteer relief and recovery organizations were particularly hard hit due to declines in financial donations and workforce loss as a result of personal and mental health concerns.

The pandemic also created unique constraints by encouraging isolation, which impeded disaster management actions that traditionally require a collective and interpersonal approach. New modes of virtual service delivery, including major technological developments such as telehealth, virtual counseling, and e-learning, became staples of routine service delivery during the pandemic. However, the sequence of weather-climate events displaced large portions of the population and disrupted the effective delivery of services that had come to rely on broadband technology. Limitations of broadband availability and the complex and continually evolving online disaster recovery assistance systems presented major challenges to households and municipalities, which created inefficiencies and vicious feedback cycles that further delayed recovery.

During a time of protracted recovery, affected households and their communities are at greater risk to the adverse effects of future disruptive events. Disaster recovery systems that traditionally associate losses from individual events were unable to process losses resulting from cumulative damages incurred during sequential disruptive events, leading to a profound effect on housing availability and scale of community recovery. Buildings and infrastructure remained in disrepair and were then further damaged in subsequent events, creating a vicious reinforcing feedback loop between household and community recovery processes. With disaster recovery funds delayed, the flexibility and "pre-positioning" of ARPA (American Rescue Plan Act, P.L. 117-2) funds allowed some local governments to innovate and address critical gaps in public assistance programming, bringing greater stability to communities that otherwise may not have been possible. Generally, the disasters of 2020–2021 underscored how resilient, safe, and affordable housing functions as vital infrastructure to sustain education, social cohesion, and economic activity in communities and throughout the region.

5

Lessons Learned

This chapter focuses on lessons learned by the Committee on Compounding Disasters in Gulf Coast Communities, 2020–2021, about the Gulf of Mexico (GOM) region in the 2020–2021 time frame. The committee defines lessons learned as the identification of either shortcomings or successes in disaster preparation, response, or recovery modifications to implement for future events. It acknowledges that some lessons are merely recognized, some are embraced and implemented, and some are forgotten with the passage of time, but the committee includes all phases of that process in this chapter. Generally, the ability of communities, governments, and systems to identify lessons learned and to carry those lessons forward through to implementation depends on the strength of their respective adaptive capacities. The committee derived these lessons learned primarily from the information-gathering sessions in Texas, Louisiana, and Alabama. **Because the committee conducted those sessions in 2022–2023, insufficient time had passed to fully assess the incorporation of lessons learned into policies and procedures.** Panelists indicated that some adjustments had occurred, but it is not possible at this time to gauge how much further the implementation phase will go or how effective it will prove to be, particularly for compounding disasters. Based on research and experience in other communities, 4–5 years is often necessary for postrecovery modifications to be made and integrated into preexisting plans and procedures (Colten, 2012, 2020; DeRobertis, 2024).

LESSONS RECOGNIZED AND LEARNED

The committee approaches this topic of lessons learned with caution. While it is important to capture and relay these ideas, the next and essential step is to incorporate them into disaster policies, procedures, and protocols and ensure their perpetuation for future events. A point of ongoing criticism is that postevent assessments more often than not represent a compilation of "lessons recognized," not "lessons learned." Cutter and colleagues (2008, p. 603) note the distinction, writing that "debriefings after the event is over are used to identify what went right and what went wrong in the response. In reality, lessons learned are merely lessons identified [recognized]. They are commonly formulated as recommendations that may or may not be implemented." Other studies described below note the gap between recognition and implementation of lessons, specifically considering the long-term value of lessons learned. Participants in the committee's information-gathering sessions acknowledged most of these gaps.

Opportunities for Broader Adoption of Lessons Learned

More effective coordination of leadership, the creation of incentives to share duties and responsibilities, preparation of mutual aid agreements among agencies, and improved communication among agencies and between agencies and the public present opportunities to shift lessons recognized to learned (Donahue and Tuohy, 2006). This can be facilitated by tailoring lessons learned reports to multiple agencies, not just an internal audience, as a way to penetrate the silos of both agency responsibilities and technical expertise (Donahue and Tuohy, 2006). Development of a process for institutionalizing change is an essential ingredient of moving a lesson toward implementation. A regional or national repository of lessons learned reports could aid in the dissemination of findings across the jurisdictional silos of state boundaries. The committee's information-gathering sessions revealed successes in multi-jurisdictional and interagency communication and cooperation. Local, state, and federal participants also noted that room remained for improvement in more thoroughly integrating planning and response.

Impediments to Implementing Lessons Learned

In examining lessons learned and repeated failures to learn from disasters, Donahue and Tuohy (2006) note that **it is common to identify**

the same lessons postdisaster repeatedly without improving response and recovery procedures and outcomes, and that local officials, community members, and others consistently identify communication, command structure, and resource deployment as elements needing improvement. Among the gaps in the process that progresses a lesson from recognition to implementation are (1) isolating findings within separate and uncoordinated organizations, (2) not sharing the findings with those on the ground and building them into training, (3) focusing on what went wrong and not what went right, (4) not ensuring that procedures and practices are designed to facilitate flexibility, and (5) not sustaining a commitment to change long enough for it to be implemented.

Kahan and colleagues (2006) cite the erosion of a sense of urgency as memories of a past event fade as another impediment to incorporating lessons learned. Although they do not address compounding events, their concept of erosion can be replaced by an eclipse of concern with prior events as new challenges emerge. Delays in or a lack of implementation can also occur in the face of conflict between what policymakers seek to implement and what the public desires. Session panelists pointed to yet another fundamental challenge to implementing lessons learned: those lessons are frequently tied to local situations, but they must be implemented in the context of regional and even federal systems that may not be attuned to or responsive to these lessons, particularly when multiple events impose demands on different public agencies with related but distinct responsibilities. When one or more disasters overwhelm local capacities, the absence of an effective regional strategy inhibits cooperation and coordination across geographic jurisdictions, complicating documentation of a potential lesson learned.

Shifting local priorities are also among the factors contributing to the erosion of lessons learned. Burby (2006) asserts that economic priorities sometimes eclipse "safe development." In the years following Hurricane Camille's landfall along the Mississippi Gulf Coast in 1969, local policies restricted development near the shore to avert the devastating storm surge. Over time, however, an anticipated bonanza from floating casinos encouraged the permitting of facilities on the same shores that had suffered serious damage from Hurricane Katrina in 2005 (Colten and Giancarlo, 2011). Although local leaders modified public policies to reflect the lessons learned from Hurricane Camille, the lure of economic gains with the rise of casino gambling prompted changes that cast lessons learned aside. Freudenburg and colleagues (2011) recount the arrival and rise of what they call the "growth machine" (or economic development priorities) to New Orleans, and how it created conditions that amplified Hurricane Katrina's impacts. While Louisiana strengthened building codes after Hurricane Katrina (and Rita) by adopting the 2012 International Residential Code in 2013,[1] high wind design requirements were omitted and eventually the 2015 edition of this code was suspended in 2017 by executive order (IBHS, 2018). These examples come from events that occurred well before 2020, yet they illustrate the slow pace of governance and policy implementation—and even the reversal of lessons learned—that can influence the outcomes of sequential hazard events (Colten, 2005, 2009). The pace of incorporating lessons learned is a vital factor in whether changes have been implemented when a subsequent event occurs. For example, investments in structural mitigation followed several hurricanes that damaged New Orleans in 1915, 1947, and 1965. The protracted processes for approving, funding, and then constructing these features left the urban area susceptible to events that occurred years after decisions had been made to fortify the city (Colten, 2015; Horowitz, 2020). Levees approved after Hurricane Betsy in 1965 had not been fully completed when Hurricane Katrina overwhelmed the region in 2005 (Colten, 2009). There is also a tendency to prioritize restoring local economic functions immediately, bringing evacuees back to their homes, and regenerating the local tax base. Such decisions can leave safety considerations languishing (Burby, 2006). It has taken Louisiana parishes that were damaged by the floods of 2016 more than 7 years to impose restrictions on certain land uses in flood-prone areas (DeRobertis, 2024). Therefore, this study of the events of 2020–2021 may miss some of the secondary actions taken to implement lessons learned and does not capture their long-term erosion.

Connecting Professionals with Lay, Local Expertise

Lessons learned that are documented may **not incorporate local, experiential knowledge**. McEwen and colleagues (2017) focus specifically on locally based knowledge of flood risk and mitigation, or what might be considered a repository of lessons learned, which they refer to as "flood memory." They underscore that flood memory is built on collective

---

[1] The International Residential Code is a comprehensive, stand-alone residential code that establishes minimum regulations for one- and two-family dwellings and townhouses using prescriptive provisions. It is founded on broad-based principles that make possible the use of new materials and is designed using model code regulations that safeguard the public health and safety in all communities, large and small (International Code Council, 2018).

experiences, which can bolster a community's resilience when it faces extreme or unusual flood events. However, risk management and disaster recovery agencies make only modest attempts to draw on local, lay knowledge, and consequently neglect a valuable archive of expertise in flood risk and flood mitigation. The dissociation of local wisdom and external agency planning impedes the incorporation of lessons learned at the local level and the filtering of local expertise into higher levels of government agencies.

Local adaptations to extreme events provided the primary means of coping with disasters long before the creation of federal and state emergency response agencies. Colten and colleagues (2012, 2015) underscore the role of social memory in enabling community resilience. They note that local expertise existed before the creation of both civil defense and the Federal Emergency Management Agency (FEMA) and has performed when other formal programs have failed. They also suggest the need to integrate this community-based wisdom with formal disaster planning and response, and the deliberate perpetuation of lessons learned between events.

Hazard scholars often use the term *resilience* to denote the capacity to cope with disasters and extreme events and recover from the devastation they cause (Adger, 2000; Aldrich and Meyer, 2015; Colten et al., 2012; Wilbanks, 2008). Resilience is multidimensional and is built in part on a number of locally based social capabilities that fall under the rubric of social capital.[2] Adaptive capacity is a reflection, in part, of social capital and a fundamental element of resilience. Strengthening social capital, or social support, is therefore essential to fortify resilience. Comprehensive participatory planning for disasters offers several paths to strengthen social capital, including bringing in local experts at the outset of planning and preparation, which can take place between disasters, as well as during response and recovery efforts (Aldrich and Meyer, 2015). Panelists representing nongovernmental organizations (NGOs) consistently advocated for more community-guided hazard planning that tapped local expertise.

LESSONS RECOGNIZED AND LEARNED: THEMES FROM THE INFORMATION-GATHERING SESSIONS

Panelists in each of the information-gathering sessions spoke about the importance of implementing lessons learned and highlighted common

---

[2] For more information on social capital, see the Chapter 2 section entitled "Social Capital and Cohesion."

themes from the lessons learned from their experiences in 2020–2021. These themes largely echo the studies cited above in their emphasis on (1) organizational structure and agency functional abilities and coordination, including advance planning alongside spontaneous training, flexibility, and adaptability; (2) funding, both existing sources and streamlined procedures to secure postevent supplemental funds; and (3) communication, vulnerabilities, disaster recovery service delivery, and housing needs. The following section organizes panelists' lessons learned by these themes.

Organizational Learning, Training, Agency Coordination, and Flexibility

Participants spoke at length about the need for improvement in the area of cooperation and coordination among various government agencies, faith-based organizations (FBOs), community-based organizations (CBOs) and volunteers, and spoke of what they had learned in this regard during the 2020–2021 time frame. Panel members from Louisiana commented on how cooperation among the parishes and between the parishes and state/local actors was well coordinated for the storms of 2020. They spoke of brokering informal arrangements to direct mutual aid and creative problem solving, which enabled responses that at times went above and beyond the call of duty. Panelists noted the importance of improving mechanisms and procedures for coordination with and accommodation of NGOs and volunteers in these engagements to ensure their integration in the disaster response and recovery processes.

The nationwide push for tracking data and analytics to monitor and lessen the spread of COVID-19 facilitated greater coordination among local NGOs, CBOs, and various governmental sectors. **The pandemic spurred improved capacity for data-informed decision-making; more streamlined operations; enhanced coordination and communication within and across organizations; and even the ability to repurpose pandemic aid, resulting in immediate benefits.** According to several panelists, the pandemic forced some organizations to create new systems for generating and tracking data to support reporting and decision-making for the first time. These new systems ultimately led to a wealth of collective knowledge and industries designed to assist organizations in data monitoring and virtual modes of operation. The new collaborations established among government representatives, NGOs, and CBOs in response to the pandemic enhanced unity, productivity, and efficacy during ensuing hurricane responses. The capacity to track, report, and make pandemic data-informed decisions now guided disaster response and recovery in the pandemic era.

Participants emphasized the importance of training and drills that enable staff to become familiar with plans and any lessons learned from prior events and that maintain a sense of readiness. Accordingly, new trainings, including surprise mock trainings with stormwater damage and blocked roads, are on the docket in Alabama. Alabama panelists also observed that full mobilization for even a tropical storm served as a means to practice for a Category 3+ hurricane. Beyond training and drills, regular standing interactions help establish working relationships prior to disasters and facilitate effective operation during times of crisis.

Those in greatest need of assistance, including marginalized, disadvantaged, and excluded groups, and specifically elderly, poor, and immigrant populations, are often the least connected and technologically ready for a disruptive event (Li et al., 2022). In response, panelists in Harris County described how their neighborhood groups and other organizations conducted trainings with these populations on the use of Zoom, FaceTime, WhatsApp, and other communication platforms.

Learning how to work remotely and delivering services with tools such as Zoom and Microsoft Teams improved efficiency and adaptability. "It prepared us for what was to come," said one panelist. "We learned to work from home; schools and daycares were already closing, so we were learning how to build that flexibility into daily life, which helped for hurricane season—that prepared us to maintain services throughout the hurricanes." These same systems that enabled learning during the lockdowns continued to function as schools were repaired.

Panelists remained committed to maintaining plans for emergency operations, but they also stressed the need to allow flexibility when unanticipated conditions arise. This was particularly true for Winter Storm Uri in Texas and Southwest Louisiana. There were no plans for this unimagined event, and spontaneity and creative adaptations were essential. Panelists emphasized that plans and structures need to allow for creative solutions that work. In Texas and Southwest Louisiana, experiences with Winter Storm Uri prompted recommendations to plan for the unexpected by considering events not included among high-probability hazards. Panelists voiced calls for dealing with unexpected scales of familiar events in locations hit by storms that moved exceptionally slowly (Hurricane Harvey) or intensified rapidly (Hurricane Laura). Alabama panelists praised their local elected officials for stepping back, asking what was needed, and allowing those working in response and recovery to do their jobs when storms unexpectedly intensified, such as Hurricane Sally, and/or wreaked more damage than expected.

Flexibility was especially important in responding to tropical cyclones in the midst of the pandemic, underscoring its value in compounding disasters. Pandemic restrictions forced organizations across different sectors to increase agility and operational streamlining. These changes manifested in new modes of virtual and hybrid collaboration to accommodate lower-density operations (physical distancing required limiting, and in some cases prohibiting, staff from convening physically to deliver services) and COVID-19 outbreaks in physical facilities. Panelists pointed to the newfound "readiness mindset" necessitated by the pandemic's rapid evolution and ever-changing guidance, which demanded pivots on short notice and creative workarounds for new challenges. As one panelist from Mobile, Alabama, explained, "The word of the year for 2020 in our office is 'pivot.' And it has continued to be our word."

Communities also adapted services that could not be provided virtually. An NGO leader in Louisiana described how the pandemic forced their organization to close its main soup kitchen, instead using food trucks and door-to-door delivery services to continue serving meals to vulnerable populations and better preparing them for the seasonal hurricanes. One panelist noted that "COVID built our resilience. It showed us that the same things we learned surviving hurricanes could be used for pandemics and vice versa." The transitions forced by the pandemic were burdensome for many organizations, particularly those with smaller budgets and limited capacity, as well as for individuals who were resource constrained with limited connectivity or technology readiness. Yet the same sectors and systems instituted to navigate pandemic lockdowns were vital to later navigating the compounding disaster recovery process once the storms struck.

Still, the COVID-19 pandemic consumed much of the bandwidth that normally would have been directed to preparing for the hurricane season, challenging response and recovery efforts, particularly when considering physical distancing. As a panelist from Mobile County, Alabama, explained, "We're not supposed to be near each other. And the crux of . . . disaster recovery is being near each other." Fortunately, the readiness and adaptability mindset fostered in the early months of the pandemic helped organizations make up lost ground for preparations and devise creative solutions to what would be a challenging sequence of evacuation, sheltering, and recovery decisions as the first hurricanes of 2020 struck during pandemic surges. As a panelist from Southwest Louisiana offered, "I think one of the major things that was discovered was that you could do a lot with a little creativity."

Funding

Funding of emergency services and long-term recovery was a shared concern among panelists from all localities. Several funding issues relate to the provision of federal disaster funds. Panelists representing local governments repeatedly spoke of the need to accelerate access to supplemental federal funds, and Congress is considering legislation addressing this issue (Graves, 2023). Local governments have immediate needs to restore basic functions, and their contractors expect prompt payment, so that considerable cash must be on hand to launch recovery efforts, or at minimum, debris removal. Municipalities that directed any unobligated funding in their budgets toward the mounting costs of clean-up and restoration of essential services experienced difficulty covering those costs. For some local governments, the pandemic had already tapped those reserves well before the start of the 2020 hurricane season.

Expedited delivery of federal funds in the aftermath of a disaster would allow municipalities to compensate local contractors and replenish the funds expended on a timetable that would permit paying for immediate debris clearance, followed by repairs and other recovery costs. Timely payment reduces the vulnerability and exposure to a second extreme event that are associated with incomplete recovery efforts (see Chapter 4). Local government panelists suggested that modifying the formula for cost sharing to reduce the local share would also enable more rapid recovery. This type of change would particularly aid small towns or counties with limited reserves for disaster recovery, much less a rapid succession of disasters. Box 5-1 describes recent federal efforts to advance the disaster recovery process.

Communities also discovered the importance of thorough documentation of damages and response/recovery expenditures for each sequential event given the event-based accounting required for federal reimbursement. A related lesson learned in Louisiana was the need for a dedicated response/recovery fund. Directing a set amount into a special account to be held until needed could provide small communities with a cash reserve when a disaster inevitably occurs. Rapid access to funds is especially important when compounding events occur.

Lessons learned differed regarding the influx of funds during different phases of the pandemic. Panelists from communities in Louisiana

BOX 5-1
Federal Efforts to Advance the Disaster Recovery Process

Expediting Disaster Recovery Act

In November 2023, U.S. Representatives Garret Graves (Louisiana) and Stacey Plaskett (U.S. Virgin Islands) introduced legislation designed to expedite the provision of immediate disaster recovery funds to states in the aftermath of a disaster. The Expediting Disaster Recovery Act (H.R. 5774, 117th Congress, 2021–2022) would require the Federal Emergency Management Agency (FEMA) to immediately fund 10 percent of estimated grant assistance under Sections 406 (Repair, Restoration, and Replacement of Damaged Facilities) and 408 (Individual Assistance) of the Stafford Act (Robert T. Stafford Disaster Relief and Emergency Assistance Act, P.L. 100-707) to respond to disasters within 30 days of a declaration (Graves, 2023).

FEMA Announces Major Updates to Individual Assistance Program for Disaster Recovery

In January 2024, the Biden-Harris administration announced extensive revisions to FEMA's Individual Assistance Program aimed at reducing administrative barriers and accelerating the delivery of financial assistance to disaster survivors as necessary to enable faster recovery. These changes to administrative policy came following the analysis of several decades of disaster claim data and public comments solicited since 2021. According to official statements, FEMA expects the new policies to be effective for new federally declared disasters occurring on or after March 22, 2024.*

Community Disaster Resilience Zones Act

The Community Disaster Resilience Zones Act of 2022 (P.L. 117-255) amends the Stafford Act, requiring FEMA to use a natural hazard risk assessment index to identify the census tracts most at risk from the effects of natural hazards and climate change. FEMA used the National Risk Index datasets to identify the most at-risk and in-need communities to create resilience zones. The zones will provide geographic focus for financial and technical assistance from public, private, and philanthropic agencies and organizations for the planning and implementation of resilience projects that can help to reduce climate change and other natural hazard impacts. The first set of zones was announced in fall 2023; the next group will be announced fall 2024.

* See FEMA HQ-24-007 for further details.

dealing with extreme weather-climate events explained that some American Rescue Plan (P.L. 117-2) relief funds could be used indirectly to aid with longer-term recovery, especially in the areas of housing, utilities, and mental health. Participants from other places noted that restrictions on early-stage COVID-19 funds limited use of those funds for storm recovery efforts. **The "pre-positioning" of flexible pandemic relief funds provided some municipalities with unexpected resources to meet acute needs such as emergency rental and utility assistance, localization of capacity to meet acute needs in vulnerable communities, and long-term recovery support for critical sectors such as housing.**

Communication

Panelists from all six counties/parishes noted the need for regular communication among emergency responders, related public agencies, FBOs, CBOs, and NGOs before and during disruptive events. They also provided examples of different pathways for that communication. Situation awareness needs during the pandemic spurred more regular flows of data and communication across organizations, including regular multisector coordination calls, initially instituted for actors responding to the pandemic.

For the organizations themselves, coordination calls dedicated to tracking pandemic case counts had to be repurposed to triage the response to the latest weather-related disaster. A session panelist from Mobile, Alabama, described how "[NGOs] made connections with various organizations: municipalities, educational institutions, public health. They provided weekly updates . . . we had a conference every single week." This practice had immediate co-benefits for those navigating the weather-climate events that followed.

Weekly community resource calls continued in some regions even as COVID-19-related risks diminished. The calls brought fragmented groups together to share information on COVID-19, disasters, schools, and other community concerns. Panelists described applying "gray-sky" lessons to "blue-sky and peace time." One panelist summarized, "We know that we all need each other and are not trying to one-up each other; less competition, more collaboration." **Beyond the exchange of vital data, the consistent communications fostered relationships that would prove vital in confronting the compounding disasters that would follow.**

Panelists praised the new frontiers of communication that saw tremendous growth because of the pandemic. Telemedicine, distance learning, and remote work enabled the continuation of education and job responsibilities and the provision of essential health care. Church groups were able to minister to and communicate with members using faith-based group chats. The new communication tools enabled community leaders and government agencies to relay information to the public when offices had to be closed, particularly because of COVID-19. They enabled the delivery of vital medical and mental health services remotely when visits to offices were limited. And they provided a communication bridge when local newspapers, radio, and television were disrupted or unavailable to the public.

The GOM region has experience with activating special medical needs shelters with auxiliary power generators to safeguard preregistered medically high-risk patients with such ailments as COPD (chronic obstructive pulmonary disease) who are dependent on electricity to operate the medical equipment on which they rely (e.g., oxygen concentrators). Registries have been established to monitor, track, and warn these populations (see Box 5-2 for an example).

The ability to collaborate virtually within and across organizations inspired other creative ways to share information with beneficiaries and constituents. NGOs in coastal Alabama assembled listservs to streamline communications and distribute "living" community resource guides that were continually updated. NGO staff manually monitored and responded to social media posts to direct individuals to resources for and information on essential community services.

BOX 5-2
State of Texas Emergency Assistance Registry

One of the most sophisticated and comprehensive medical registry systems in the nation is the State of Texas Emergency Assistance Registry. STEAR registers people who are medically fragile; those with disabilities; and those with access and functional needs that may include communication barriers or limited mobility and therefore need additional medical, transportation, and personal care support during an emergency event. Together with structural reforms, such registries can help mitigate the inequities in disaster preparation, response, mitigation, and recovery for medically high-risk patients.

When possible, some organizations distributed physical cards to remind residents to call 2-1-1 for essential community services, and even mobilized word-of-mouth campaigns by planting key messages with well-connected intermediaries in communities. SMS (Short Message Service) notification systems and phone trees were developed and used to accommodate people without access to smartphones or data plans that could use providers' apps. As one Louisiana NGO leader remarked, "There was not a sense of competition. There was just a huge amount of communication because we know each other at a community level."

Related to risk communication, but not in terms of technology or language, public reliance on both traditional media (newspapers and television/radio news) has declined, and disinformation and misinformation are more pervasive and influential today than ever before (Lipka and Shearer, 2023). Public officials and community leaders observed that misinformation made their jobs more difficult and contributed to frustration and erosion among trust of those in need of assistance. While social media played a vital role in disseminating information about impending weather-climate events and response efforts, these outlets also provided a medium for the distribution of misleading and inaccurate information. The erosion of trust among public health and disaster professionals and the communities they serve was evident with information about COVID-19, which in turn led to skepticism toward authorities in the wake of other weather-climate events. Absent the COVID-19 pandemic, this situation might not have occurred. Panelists recommended the development and implementation of processes for contending with misinformation and disinformation such as trust building and participatory planning.

Disaster Recovery Service Delivery

The importance of equitable delivery of services to affected communities emerged as a powerful theme in the panel discussions on lessons learned. Some panelists in each session voiced the need to incorporate equity into planning for and recovery from future disasters. There was general consensus that vulnerable communities—communities of color; low-income, medically vulnerable, and elderly populations; those experiencing homelessness; and mentally and physically challenged individuals—fared the worst during extreme events, and that response and recovery programs needed to take this into account. "We have to lift the bottom up," summarized one panelist. Another panelist from Alabama noted that conversations about inequities are becoming more inclusive.

A refrain heard more than once from panelists was: "The greedy are first in line, the needy are at the end of the line." Some of those directly providing services were convinced that the most vulnerable are least able to position themselves to take advantage of public and civil society services after a disaster, while those with the ability to access disaster aid facilities and websites are likely to get more than their fair share. This situation demands tools to ensure that people without transportation, phone/internet service, technological literacy, or language skills are not excluded or neglected. Local NGO and CBO participants repeatedly commented that residents with the greatest needs find themselves falling further behind in the wake of a disaster. This situation was particularly evident in 2020–2021 when COVID-19 forced virtual interactions that made it more difficult for those without computer access or skills to apply for aid. As the pandemic spread and compounded with other disasters, it tended to amplify these inequities. Numerous comments from panelists affirmed the academic research on the inequitable provision of relief aid to marginalized communities.

Panelists expressed the view that government bodies need to better incorporate FBOs into disaster response. They also stressed the need for governments to develop better protocols for reducing postevent scammers from preying on desperate homeowners. While these concerns exist for all disaster responses, they become more acute when compounding disasters occur, a time when increased local capacity is needed and household vulnerability is heightened.

There was an abundance, even an excess, of food available in Lake Charles after Hurricanes Laura and Delta. Some panelists suggested that other resources might have been more beneficial. Local leaders recommended establishing a management center for both internal and external aid groups to direct incoming donations toward unmet needs and allocate resources to the most vulnerable. Generous, yet ill-suited, donations such as excessive food complicate the role of NGOs in all circumstances and can be an unnecessary demand on staff in the increasingly complicated conditions of compounding disasters.

NGO panelists noted the protracted process involved in securing funding for disaster relief and recovery, and that this support is tied to specific events. They suggested that FEMA issue new clear, concise, and intuitive policies and procedures for requesting and receiving individual assistance, and that the timing for appropriations for housing assistance to state governments be updated. By the time funding arrives, said one panelist, "many people have already left." Government officials agreed with the timing challenges, adding that it was difficult to fund contractor and debris removal services in sequential disasters while awaiting FEMA reimbursements from a prior disaster.

Housing Needs

Housing is one of the most critical needs of those displaced by disruptive events (see Chapter 4). Since construction workers and supplies may not be able to meet the demand for adequate shelters in a timely way, families may be unable to complete repairs before temporary housing expires. As a Louisiana state official explained, "There's just so much of a gap between where the response assistance ends and where recoveries begin." State and federal programs provide temporary shelter, but local officials recommended adding transitional, longer-term sheltering options and advanced planning of staged temporary housing. Additionally, given the lack of public transportation options in many rural communities and even small cities, they suggested that aid should extend to those who have lost not just houses but also cars, so they are able to maintain their jobs. Local officials noted further that they intended to eliminate future virtual home and building inspections to ensure that their residents will receive accurate damage assessments and avoid the complications, costs, and delays caused by inadequate and inaccurate remote inspections conducted during 2020–2021.

Efforts to design and build sustainable, storm-resistant, and prefabricated housing were recommended after Hurricanes Katrina and Rita. Few examples of such housing were built, however (Alter, 2021). Integrating community guidance and priorities can facilitate better outcomes for temporary and replacement housing. Durable but quickly erected homes could both provide temporary poststorm shelter, serve as the core for long-term dwellings, and begin to address affordable housing and social isolation challenges (Alter, 2021; Moser Design Group, n.d.). This is one approach that could reduce the need for trailers as temporary and disposable shelters. A design competition might yield creative ideas for cost-effective homes that would provide quick and safe shelters while helping to retain local residents and the essential labor pool in the longer term.

LESSONS RECOGNIZED AND LEARNED: NONGOVERNMENTAL ORGANIZATIONS, COMMUNITY-BASED ORGANIZATIONS, AND GOVERNMENT OFFICIALS

The information-gathering sessions drew on the expertise of officials and community leaders with different levels of authority, responsibility, and funding. The discussions during the sessions therefore highlighted not only the shared lessons learned summarized above but also distinct sector perspectives, as discussed below. All sessions exposed a keen awareness of the compounding impacts of the COVID-19 pandemic on staffing and emergency response.

NGOs and CBOs

Panelists representing NGOs and CBOs tended to be most attentive to vulnerable communities. Poverty and job interruptions were identified as critical impediments to recovery among those served by NGOs. NGO representatives emphasized the critical importance of community participation in planning and preparation for hazard events. Panelists cited communication through social networks—both in person and via social media—as a critical tool at the local level. Participants observed that funding, especially for housing, was slow to arrive and therefore delayed recovery. Overall, these organizations found strength in their communities and advocated for greater coordination with government bodies. Generally, panelists recognized the need to ensure that the most impoverished and least well-equipped to recover from compounding disasters are prioritized in mitigation investments.

Government Officials

Government officials tended to have a broader view than that of NGO panelists and often spoke about the regional scope of disasters. A common theme revolved around improvements needed in FEMA's approach to valuing damages and its provision of temporary housing after a disaster. Government officials from Texas and Louisiana pointed out the need to improve supply chains to ensure efficient delivery of essential supplies during and following a disaster, which would demand greater regional and national coordination. Many panelists mentioned examples of improvements in interagency communication as lessons learned and implemented, but these panelists also recommended further refinements in communication among all levels of government, including better and more efficient localized data to prioritize recovery spending. Local and county/parish officials were particularly cognizant of impacts on local economies and recoveries due to the long-term departures of residents. They also expressed concern about the erosion of trust toward government officials and the need to address misinformation.

Some officials observed that recognition of social and economic inequities had improved but that uneven delivery of aid and assistance in the wake of disasters had nonetheless continued. They cited the need to establish recovery goals and plans targeting resilient and equitable outcomes. However, representatives of more sparsely populated jurisdictions pointed to inadequate numbers of emergency management staff in the wake of compounding disasters, with ramifications for the effective implementation of preparation and recovery plans and achievement of goals. Box 5-3 gives examples of GOM states' after-action reports and the role they play following a disaster.

The information-gathering sessions highlighted the need to build equity and climate change into the fabric of the disaster planning and response enterprise, particularly regarding disaster mitigation. Panelists stressed that addressing equity must begin with deliberate, comprehensive participatory planning that elevates community concerns *before* a disaster and not in its wake and emphasized that this type of planning must include frontline communities. Participatory planning processes open up space to discuss goals, values, adaptation and mitigation strategies, local observations, resources, and priorities. They are facilitated by providing convenient and/or multiple meeting times; offering language translation; and supplying transportation, food, and childcare (USGCRP, 2023).

LESSONS IMPLEMENTED

Despite shortcomings in implementation of lessons learned, communities in the GOM region have seen examples of improvement. Donahue and Tuohy (2006) assert that it is essential to identify successes and strive to repeat them in tandem with making improvements. If this was true in the days following Hurricane Katrina, it is even more pertinent as the GOM region faces the effects of compounding disasters.

BOX 5-3
After-Action Reports: Addressing Unmet Needs and Documenting Lessons Learned

After-action reports are prepared by state and local agencies in the wake of a disruptive event. After-action reports include unmet needs assessments and detail how disaster recovery funds will be allocated to address remaining long-term recovery unmet needs. Additionally, the reports document the actions taken during the response and coordination of efforts among local, state, federal, and nongovernmental partners following a disruptive event and the results of those actions. They aim to identify shortcomings, unmet needs, and successes in the actions taken by governmental and community-based organizations to overcome the impacts of disruptive events, compare desired and actual outcomes, and document specific improvements needed for future incident responses. Serving as formal documentation of lessons learned, they typically report on lessons from a singular event rather than multiple, sequential events and are a common feature of disaster response and recovery actions and have been for decades.

Louisiana

Following Hurricanes Laura and Delta in 2020 and Hurricane Ida in 2021, the state of Louisiana prepared a Proposed Master Action Plan (State of Louisiana, 2022) that details how funds will be allocated to address unmet needs. Priorities identified included housing and public infrastructure. Unmet needs related to housing included rehabilitation and reconstruction of damaged housing, construction/rehabilitation of affordable housing, temporary rental assistance, second mortgages for gap financing, and assistance with flood insurance payment. The state secured a hazard mitigation grant to address these needs.

Among the infrastructure needs identified, the after-action report notes that disaster plans rely heavily on surface transportation networks that can often be inundated by flooding. Submerged roads can greatly impede rescue efforts and delivery of disaster services to those in need. In response to this identified unmet need and lessons learned following the 2016 and 2018 floods in Louisiana, the state designed the Watershed Initiative (State of Louisiana, 2023). This is not a report, but a strategic approach to coordinate funding, data, and resources among five state agencies to reduce flood risk. This initiative represents a framework for planning and executing preparedness and mitigation procedures to address unmet needs and reduce flood risk. Funds from the state's Community Development Block Grant Mitigation funds are being targeted to address some of the identified infrastructure needs.

Texas

Following Winter Storm Uri in 2021, the city of Houston prepared a recovery action plan (City of Houston, 2022), which included an assessment of unmet needs. As with Louisiana's unmet need priorities, officials identified housing and public infrastructure, but they also noted the critical role of community lifelines. The report specifically addressed the compounding impacts from previous disasters and acknowledged that they contributed to unmet needs. It noted that climate change may facilitate a storm comparable to Uri that might exceed the capacity of homes to protect people from such extreme temperatures. Houston's housing stock was vulnerable to the freeze because it had yet to be repaired after the floods in 2015 and 2016, Hurricane Harvey (in 2017), and Tropical Storm Imelda (in 2019), since many homes were un- or underinsured. The report noted further that Federal Emergency Management Agency (FEMA) funds tend to go to homeowners or families living in single-family homes, and that this likely leads to undercounting and undervaluing the damage from disasters, rendering data insufficient for analyzing the true impacts of these events. The storm followed previous patterns where outstanding needs not covered by insurance were in majority-Black and majority-Hispanic zip codes. There were notable concentrations of housing that suffered "compounding damages" (City of Houston, 2022, pp. 20–21), with impacts from multiple hazard events in areas with high social vulnerability. Some panelists cited improved insurance coverage as one way to accelerate recovery.

Loss of power during Winter Storm Uri disrupted normal planning for and response to an extreme weather event. Critical infrastructure, particularly the water system, experienced disruptions and recovery challenges due to power loss. As a reflection of the after-action report, the city of Houston submitted requests for hazard mitigation grants for emergency generators for both its water delivery system and its emergency services (police and fire). Additional generators are also needed for drinking water and wastewater operations, traffic and stormwater controls, and emergency services. Houston's mitigation plan focuses primarily on flood and wind risk and not winter storms. Uri proved that a winter storm can have dramatic impacts that require improvements in community lifelines such as transportation, essential services, health care, communications, and energy.

continued

BOX 5-3 Continued

Alabama

In the wake of Hurricanes Sally and Zeta in 2020, the state of Alabama prepared an action plan that identified unmet needs (Alabama Department of Economic and Community Affairs, 2022). The report also noted the compounding impacts of the COVID-19 pandemic. Its aim was to mitigate the impacts of future disasters in accordance with the U.S. Department of Housing and Urban Development's Community Development Block Grants Mitigation Program. Outstanding needs include housing and public infrastructure.

Hurricane Sally exacerbated existing housing shortages in Mobile and Baldwin Counties, especially for low-income households. Housing shortages after the storm were most acute in the rental market and affordable housing. In Mobile County, the majority of applicants for FEMA assistance were renters and did not have flood insurance, making unmet needs due to flood damage especially pronounced there. The Home Recovery Alabama Program prioritized these types of properties.

After the 2020 storms caused damage to housing stock, roads, bridges, wastewater and public water facilities, public buildings, recreational facilities, and public utilities, the state's report identified numerous unmet mitigation needs for public buildings and housing. These include retrofitting of public buildings to withstand severe storms (tropical cyclones); improvements in stormwater management; and security, elevation, and relocation options for coastal housing in advance of sea level rise and storm surge.

Innovations: Lessons Learned and Implemented Spontaneously

As described in this report, spontaneous improvisations are a common ingredient of disaster response. It is folly, however, to rely on disaster to be the mother of invention. Documentation and effective distribution of lessons learned after disasters and the subsequent incorporation of these lessons into enduring protocols and practices is an essential building block of adaptive capacity. Moreover, the committee contends that even communities that consider themselves to be "resilient" must not be complacent or expect to endure future events without systematic investment in adaptive capacity at the community level.

Session panelists made note of some lessons learned and implemented spontaneously in the face of the unprecedented nature of disruptive events in 2020–2021. In Houston, officials realized that providing residents with water shutoff tools and basic instructions helped them shut off water during Winter Storm Uri, thereby preventing ruptured pipes and preserving water pressure citywide—a spontaneous lesson implemented in the midst of crisis. Some NGOs in Texas developed a system that allowed residents to apply for and quickly receive gift cards so they could purchase necessary supplies in the wake of a disaster and before they were able to receive aid from FEMA. Although the gift card distribution system was overwhelmed when first implemented after Hurricane Harvey in 2017, it was revamped and operated successfully for Winter Storm Uri. In another example, the provision of non-congregate shelter in areas hit by tropical cyclones was an adaptation of necessity put into motion as storms approached coastal areas in the midst of COVID-19 social distancing restrictions. Prearranged agreements with hotels that were eager for business because of the decline in travel during the pandemic accelerated sheltering of evacuees.

Using buses as warming shelters was a novel tool for Harris County during Winter Storm Uri. While municipal buses have been used as temporary shelters and cooling centers in the GOM region, their application for warming needy residents and charging cell phones was a lesson learned put into action during Uri.

One panelist, echoed by others, described how, in combination with social distancing recommendations and the prompt return of residents to Cameron and Calcasieu Parishes, where nearly all the buildings were damaged by Hurricane Laura, parking lots served as important infrastructure and community gathering locations. Residents constructed a makeshift gas station in the parking lot of a Dollar General store, where food, ice, and gas were available for anyone who needed them. The parking lot became the "territory of hurricane response," one panelist summed up, where residents could grill, barbecue, and share what would quickly spoil in their freezers because of power losses. One panelist recalled that crawfish farmers trucked in more than 100,000 pounds of ice, where volunteers passed out multiple "18-wheeler[s] [of ice] in 2 days with a shovel and a 5-gallon bucket." A creative Boy Scout project involved placing solar power charging stations made from plywood boxes on street corners. People left their phones at stations to be charged without fear of theft. Public septic services were set up around communities that otherwise lacked water or wastewater facilities.

Lessons Learned and Implemented from Events Prior to 2020–2021

Lessons learned from prior events that proved useful included the importance of advance staging of emergency equipment and supplies. "Go-bags"—hurricane emergency kits packed with essentials including rechargeable battery packs and plastic bags in which to seal critical documents—were pre-positioned for swift distribution in the event of an emergency evacuation. Pre-positioning of other essential items throughout a city or county/parish enabled speedy distribution even when debris or iced roadways impeded movement of supplies. Baldwin County, Alabama, established advance contracts with debris removal contractors and maintenance agreements with the state for debris staging areas following the accumulation of an "unreal amount" of debris after Hurricane Sally. Plans to make advance arrangements for house inspections in Southwest Louisiana are a priority there. Advance arrangements to secure engineers with expertise in preservation of historic structures also have been put in place so these structures can be more swiftly repaired after a disaster. Parish officials mentioned that they now establish predisaster contracts for debris removal and home repair with contractors who are familiar with the architectural style of historic buildings to expedite the restoration process.

Officials in Southwest Louisiana implemented a number of lessons learned from Hurricane Rita that paid dividends during Hurricane Laura. For example, local officials permitted residents to return home almost immediately after Hurricane Laura so they could assess damage and begin repairs. As a parish official said half-jokingly, "I think we would've had a mini revolt from people if you told them not to come back. Plus, you can't stop them, they know the roads." This course of action was intended to address the challenges faced by residents who were not allowed to return for lengthy periods after Hurricane Rita in 2005. Its implementation was preceded by an early evacuation to ensure that residents had time to depart the region safely (evacuation timing has been enhanced by improved hurricane forecasting), although residents were reluctant to evacuate based on their experience in Rita, which put lives at risk. However, there was a trade-off arising from the rapid return of residents. With rapid return, road clearance crews, utility repair teams, and other emergency response activities were hampered by the higher volume of traffic due to returning residents. In addition, this traffic placed greater demands on limited fuel supplies. Vehicles that ran out of gas or had flat tires due to road debris further hampered efficient response activities. But a benefit of the policy change was clear when residents readily evacuated for Hurricane Delta with the knowledge that they would be allowed to return early.

Officials modified building codes following Hurricanes Katrina and Rita in 2005. Panelists reported that many of the newly constructed buildings survived Hurricane Laura. **The call for resilience to future climate-related disasters echoes national agendas and also emphasizes the importance of proactive risk mitigation.**[3] One panelist noted that "We're really good at response, we're really good at recovery . . . but we do need to start talking about mitigation first and figuring out how to fund the mitigation first. We don't all start at the same place in a crisis. If we don't figure out how to get the baseline up . . . then we will not be able to be resilient as a community." This is a significant development in a region where there is considerable skepticism toward climate change among elected officials and decision-makers.

Cameron Parish buried its broadband cables after Hurricane Rita and reported that this service survived Hurricanes Laura and Delta. The 2-1-1 program, an emergency call center hotline established after Hurricane Rita to connect residents with active aid organizations that could assist with specific needs, proved effective following the 2020 storms. The program is based in Monroe, Louisiana, which is less susceptible than coastal Louisiana to hurricanes.

Following Rita, residents learned how to be better prepared for the recovery process. Their disaster kits with water, canned goods, crank radios, and other essential supplies now included tire plug kits to contend with the widespread debris on the roads. One participant explained, "Nobody wants to spend money to be prepared. Then, after Rita, people started to understand the need to save money and be prepared for hurricane seasons, just like you saved for vacation." More people now own backup generators, for example. In Louisiana, Cameron Parish (population 6,000) found that it was essential to have a full-time paid staff person with duties for managing emergency operations. Fire and police personnel, who commonly double as emergency managers in smaller communities, have other responsibilities and require a higher level of expertise needed to manage disaster responses.

Drawing on neighborhood resources and programs also aided residents who were unable to evacuate. Neighborhood hub houses in Houston

---

[3] The term *mitigation* is used in the disaster risk field to describe improvements to structures and systems to improve their resistance to hazards. The term *adaptation* is used for such actions when discussing climate change, as the term mitigation is reserved in this field for actions taken to reduce greenhouse gas emissions.

assemble essential provisions before events and provide water, both non-perishable and hot food, and other critical supplies to residents in certain neighborhoods. This concept is similar to the neighborhood "lighthouses" an NGO network called Together New Orleans is setting up in the wake of Hurricane Ida (Lakhani, 2023). The community lighthouse network consists of solar-powered disaster response hubs pre-positioned with emergency supplies, air conditioning, and battery power for use in the wake of power outages caused by disasters large and small; solar and battery power are also integrated into churches and schools for that same purpose (NPR, 2023).

Lessons Learned over Time: Building Codes

Table 5-1 chronicles the revisions to building code adoption and enforcement as captured by the Insurance Institute for Business and Home Safety (IBHS) *Rating the States* reports from 2012 to 2024. The 2017–2018 scores, shaded in yellow, best encapsulate the code environment that was governing new construction at the start of the period covered by the study. However, these code environments can take years to affect practices or address the vulnerability of the building inventory, given the limited number of new buildings constructed each year. The new 2024 report (IBHS, 2024) may provide evidence of lessons implemented following the study period. For example, Louisiana is the highest-performing state in building code adoption and enforcement among the GOM states, and second only to Florida in the region, increased its overall score 10 percent between 2018 and 2024 by adopting the latest International Residential Code (2018) and mandating training of officials.

Rating the States now recognizes Mississippi as the most improved state, progressing from its initial rating of 4 (lowest ever) to 44 in the most recent evaluation cycle. The state now requires statewide licensing of contractors, though it remains deficient in other areas, particularly due to the ability of jurisdictions to opt out of the building code adopted in 2014. Texas and Alabama similarly remain among the lowest-performing states because of the lack of statewide code adoption and enforcement, even slightly backsliding after the study period. Alabama's attempt to pass a statewide building code in 2023 failed in the state legislature, despite the recent experience with Hurricane Sally. Despite the deficiencies at the state level, 16 of the 24 permitting jurisdictions within Mobile and Baldwin Counties enforce the Coastal Construction Code Supplement, created after Hurricanes Ivan (in 2004) and Katrina (in 2005), and based on the IBHS FORTIFIED Home

TABLE 5-1 Insurance Institute for Business and Home Safety *Rating the States* Scores for Alabama, Florida, Louisiana, Mississippi, and Texas, 2012–2024

Alabama	2012	2015	2018	2021	2024
Code adoption	0	8	9	8	7
Enforcement officials training	0	0	0	0	0
Contractor licensing	18	18	18	22	22
Total	**18**	**26**	**27**	**30**	**29**
Florida	**2011**	**2014**	**2017**	**2020**	**2024**
Code adoption	48	48	49	49	49
Enforcement officials training	22	21	21	21	21
Contractor licensing	25	25	25	25	25
Total	**95**	**94**	**95**	**95**	**95**
Louisiana	**2012**	**2015**	**2018**	**2021**	**2024**
Code adoption	48	46	47	46	49
Enforcement officials training	16	14	14	14	20
Contractor licensing	10	22	22	22	22
Total	**74**	**82**	**83**	**82**	**91**
Mississippi	**2012**	**2015**	**2018**	**2021**	**2024**
Code adoption	0	24	23	24	24
Enforcement officials training	0	0	0	0	0
Contractor licensing	4	4	5	5	20
Total	**4**	**28**	**28**	**29**	**44**
Texas	**2012**	**2015**	**2018**	**2021**	**2024**
Code adoption	18	22	19	19	18
Enforcement officials training	0	0	0	0	0
Contractor licensing	0	14	15	15	15
Total	**18**	**36**	**34**	**34**	**33**

NOTE: IBHS's *Rating the States* scoring system evaluates 47 components of building code adoption, enforcement, licensing, and education to assess the effectiveness of a state's code program. States are assigned a composite score on a 0 to 100 scale, where 100 represents the highest possible levels of code adoption and enforcement. The yellow-shaded column is the year treated as the reference frame preceding the study period; the green-shaded column is the year treated as the frame for potential lessons implemented following the study period.

SOURCE: Component scores provided by IBHS derived from the five IBHS reports from 2012 to 2024.

BOX 5-4
Strengthening Construction Standards

Despite their deficiencies in adopting and enforcing building codes statewide, some Gulf of Mexico states have been successful in reducing housing vulnerabilities through other mechanisms. Coastal communities in Alabama have worked to raise construction standards at the county level. For example, 16 of the 24 permitting jurisdictions within Mobile and Baldwin Counties enforce the Coastal Construction Code Supplement created after Hurricanes Ivan (in 2004) and Katrina (in 2005), based on the Insurance Institute for Business and Home Safety, or IBHS, FORTIFIED Home standard (SHA, 2021). This, along with the Strengthen Alabama Homes grant program, has led to a growth in homes with FORTIFIED designations along the Alabama coast, approximately 95 percent of which incurred little to no damage (IBHS, 2021) from Hurricane Sally. Meanwhile, the Texas Windstorm Insurance Association enforces code and inspection requirements for those seeking windstorm and hail insurance through the state's wind pool (IBHS, 2021).

---

standard. Alabama is now home to the most successful voluntary windstorm mitigation program in the United States. See Box 5-4 for successful efforts to reduce housing vulnerabilities in the GOM states.

LESSONS LOST

Several examples of lessons lost or at least misplaced also emerged in the information-gathering discussions. GOM residents pride themselves on their awareness of hurricane risks and their capacity to cope with these storms. Nonetheless, the impact of a "generational" storm such as Hurricane Rita in 2005 led some residents of the region to assume there would not be another storm of that magnitude in their lifetime (before Hurricane Rita, the generational storm was Hurricane Audrey in 1957). As a public official in Cameron Parish stated, "We always prepared for hurricane season, but I just think there was this thought that we're not going to see another Rita in our lifetimes. . . . And we were wrong." Consequently, many let their guard down in advance of Hurricanes Laura and Ida.

In addition, overall planning and preparations did not adequately address exceptional events such as Winter Storm Uri, despite a severe Texas freeze in 2011 that caused power outages. Panelists mentioned that one exception in Texas occurred in El Paso, a city not in the GOM region, but on the border of New Mexico and Mexico, where the electric company opted to winterize its equipment following 2011's freeze. Residents there did not experience the extensive or damaging power outages (Busby et al., 2021) suffered by many other Texas residents.[4]

Panelists described another lesson lost when rebuilding and issuing of permits for new construction in at-risk areas continued following floods. A Texas state official spoke at length about how "we rebuild the same buildings in the same places. Prices are going to rise because of supplies, but also because of disasters and stronger, wider storms. We need to tie resilience to recovery. We need to *build back more resilient*," he emphasized. "I am tired of the phrase, 'We will rebuild.'" After a flood, it was suggested, officials should determine how elevated houses and the surrounding developments fared, for example, and then adjust as needed. Another official on this panel replied that "it's easy to provide federal money, but it's the idea of *how* to build, rebuild, and coordinate that is hard." This pattern reflects the loss of flood memory and lessons lost. See Box 5-5 for information on hazard mitigation.

The commitment to risk mitigation wanes as more time passes after a disaster. Louisiana panelists noted the pattern of avoiding spending to prepare. Disaster relief funds available from the federal government also contribute to deferring mitigation investments because of agency budget demands for relief programs.

Many of the lessons lost from prior events have been set aside or neglected. In some cases, lessons are just rediscovered. The repeated recognition of lessons learned underscores the tendency among hazard managers and society in general to allow for an erosion of readiness and a loss of the sense of urgency, and highlights the importance of taking steps

---

[4] The Texas State Comptroller (2021) reports several steps taken to address the power delivery crisis that resulted from Winter Storm Uri: The legislature has taken action to (1) revamp the Electric Reliability Council of Texas and require all members to be Texas residents; (2) require the creation of a statewide alert system to notify residents when the power supply is in jeopardy; (3) direct the Texas Energy Reliability Council to ensure that energy industries address human needs and critical infrastructure; (4) prepare a "supply chain map" to help better respond to shortages of critical supplies; (5) require winterization of energy facilities; and (6) modify the electric market structure so that customer costs are not inflated during a crisis.

BOX 5-5
Investing to Withstand Extremes

The merit of hazard mitigation is widely known, yet it is often not prioritized. In Louisiana, for example, on average, residents suffer nearly $260 in direct disaster losses for every $1 invested in mitigation, while the federal government spends $10 on Louisiana's recovery for every $1 invested in mitigation (Gall and Friedland, 2020). These mitigation investments are wholly inadequate to counter the state's damage losses. The Multi-Hazard Mitigation Council (2019) evaluated a broad suite of mitigation measures and determined that federal mitigation grants save $6 for every $1 spent, while adopting minimum code requirements saves $11 for every $1 spent.

With the many disasters that befall the Gulf of Mexico region, lessons *can* be learned and implemented. Building to withstand knowable extreme events allows families to evacuate but return back to a sound, durable home following the event. Whether through modern building codes that drive new construction practices or through grant dollars infused into low-/middle-income housing development and infrastructure, a better built environment can be the new norm. While federal agencies can nudge building codes through grant guidance, only states and local communities can adopt and enforce them. The small increase in construction costs to meet modern building codes through structures capable of withstanding the impact of extreme events pales in comparison with the direct—and indirect—costs of rebuilding.

to perpetuate and institutionalize lessons learned. Box 5-6 offers an example of a failure to perpetuate lessons learned in Louisiana.

SUMMARY OF KEY FINDINGS

Insufficient time has passed since the compounding disasters of 2020–2021 and the committee's subsequent information-gathering process to fully assess the incorporation of lessons learned from those disasters into policies and procedures. Through its review of after-action reports from disasters that occurred prior to the time frame of the Statement of Task and its qualitative information-gathering process, it is clear that lessons

BOX 5-6
Example of a Failure to Implement Lessons Learned

An after-action report following Hurricane Isaac (in 2012) identified one of the failures in a parish near New Orleans as the decision of emergency teams not to consult the procedural manual (Louisiana Governor's Office of Homeland Security and Emergency Preparedness, 2012), which had been compiled in the wake of Katrina (in 2005) to contend with future hurricanes. This is an example of the failure to perpetuate lessons learned, which regular drills and training exercises could have prevented by reminding emergency professionals of these protocols. Moreover, personnel turnover and unsynchronized revisions to disaster plans can allow for the erosion of institutional knowledge.

learned remain in various states of adoption and implementation. While it is common to identify and document lessons learned through a disaster recovery process, it is also common to identify the same lessons repeatedly over time without improving planning, response, and recovery procedures to improve disaster outcomes.

The evidence presented to the committee makes clear that robust adaptive capacity is required for communities, governments, and systems to first identify lessons learned and to then carry them forward toward implementation. A key finding by the committee is that the COVID-19 pandemic spurred rapid innovation and improved capacity in some areas, including the increased ability for data-informed decision-making, streamlined operations, enhanced coordination and communication within and across organizations, and the ability to repurpose federal financial assistance to meet local needs. All of these innovations resulted in immediate benefits among community members and strengthened the relationships necessary for improved response to future disruptive events and disasters. Notably, the pre-positioning of flexible pandemic relief funds provided some municipalities with unexpected financial resources to meet acute needs of community members, such as emergency rental and utility assistance, and public health services. However, these funds are not indefinite, and many communities will scale back or discontinue this surge of activity as federal authorizations expire and financial resources are depleted.

The committee's information-gathering sessions also underscored the need to intentionally incorporate equity and climate change into the fabric of disaster management, particularly mitigation efforts that seek to reduce both community exposure and long-standing vulnerabilities, including residential building codes, construction standards, and affordable housing.

6

Conclusions: Reducing Compounding Disaster Risk by Addressing Vulnerabilities and Exposure and Building Adaptive Capacities

Amid increased climate-related stress on people, communities, and the systems that support them, the risk of disasters and other major disruptive events continue to increase, along with the likelihood that more communities will experience the debilitating effects of compounding disasters. The physical, mental, and emotional effects of the overlapping response and recovery phases of sequential disasters compound and impose increased demands on organizations, agencies, and households. Consecutive and overlapping disruptive events, such as those experienced in the Gulf of Mexico (GOM) region in 2020–2021, can stress households, community-based organizations (CBOs), and local authorities beyond their capabilities, which in some cases may not be sufficient to contend with a single shock.

The insidious and complex effects of climate change require a new approach to understanding, planning for, and adapting to disaster risk across the systems that underpin the livability of all communities. The Fifth National Climate Assessment shows that current adaptation measures are insufficient to reduce climate-induced disaster risk, and that more efforts aimed at transforming systems are needed (USGCRP, 2023). Historically, disaster management in the United States has reacted to the occurrence of individual events. Disaster recovery processes and the funds they make available offer a potential opportunity to address vulnerabilities and reduce the impact of future disaster outcomes. However, the benefits of mobilized recovery funding rarely manifest among the vulnerable. Instead, current disaster recovery processes, too often, exacerbate existing vulnerabilities.

These inequitable effects place additional burdens on communities that continually stress the capacity of their struggling support systems. Moreover, the underlying physical and social vulnerabilities within communities continue to increase. These vulnerabilities increase the likelihood that minor disruptive events will become localized disasters. Unmitigated, compounding disasters threaten to destabilize communities throughout the GOM region.

The Committee on Compounding Disasters in Gulf Coast Communities, 2020–2021, recognizes that transdisciplinary research and applied science in disaster resilience continues to advance rapidly in the wake of recent major events in the United States and around the world. This report is intended not only to contribute to the evidence base being generated by these efforts but also to support calls for community-guided action aimed at reducing systemic vulnerabilities, mitigating risk, and increasing adaptive capacities that enable communities to handle the complex challenges that confront them today and those likely to arise in a more climate-uncertain tomorrow.

As introduced in Chapter 1 and discussed throughout this report, the reduction of compounding risk requires increased understanding of complex interactions of "meshed" exposure and vulnerability variables that determine a community's sensitivity to experience disaster and intentional action to improve baseline conditions. The ability to make these improvements is predicated on increasing adaptive capacity, which is in turn dependent on the ability of institutions, organizations, governments, and the communities they serve to access and effectively mobilize social, scientific, technical, administrative, and financial resources across systems and functions that underpin communities.

The committee's research, analysis, and deliberations yielded a number of conclusions regarding the disruptive events that occurred in the GOM region in the 2020–2021 time frame. Grounded in the evidence base and illuminated by the profound experiences relayed to the committee by the affected communities during the course of this study, the conclusions presented here underscore the need to reimagine efforts that support disaster preparedness, mitigation, and recovery in an era of intensifying weather-climate hazards and increased risk of compounding disasters. In this era of compounding disasters, all entities in the disaster management enterprise need to redouble efforts to support communities to prepare for, respond to, and recover from each disruptive event that ensures equity, fairness, and justice. Conclusions 1–3 address the expanding realities of compounding hazard risk; conclusions 4–6 identify the need for improving and expanding our understanding of the new scale and temporal scope of compounding disasters; conclusions 7–9 refer to the need to bolster public health, mental health, and community-based organizations and their adaptive capacity; and conclusions 10–14 identify new approaches to contending with compounding disasters.

The Expanding Realities of Compounding Disaster Risk

Conclusion 1: Compounding disasters introduce new, interconnected, and complex risk scenarios due to the increased potential of multiple hazards overlapping in time and space.

As established in Chapter 2, recent studies in attribution science show that climate change is causing an increase in the frequency and/or severity of tropical storms, heavy rainfall, and extreme temperatures. The intensification of these and other more extreme weather-climate events are intersecting with areas experiencing high levels of health disparities, social vulnerabilities, and increased exposure due to population growth in hazard-prone areas. This convergence is resulting in prolonged and overlapping periods of disaster recovery. While hazards may be unalterable, their impact can be reduced by collectively building adaptive capacity—whether preemptive, in real time, or through the recovery process—within vulnerable and exposed systems, institutions, and people.

Conclusion 2: Increased compounding disaster risk requires communities to plan and prepare for the co-occurrence of multiple and varied disruptive events that interact with societal exposure and vulnerabilities to amplify overall disaster impact.

Two phenomena of global scope and scale—the sudden emergence and ongoing evolution of the COVID-19 pandemic and climate change—fundamentally influenced concurrent disaster events more circumscribed in time and place. The evidence presented in Chapter 3 made clear that while GOM communities have plans in place for addressing more common climate hazards (i.e., hurricanes), the same level of planning did not exist for less common events (i.e., Winter Storm Uri, the COVID-19 pandemic), let alone the co-occurrence of multiple events. Advancing adaptive capacity involves understanding how disaster events impact specific localities and funding locally led, long-term planning for risk reduction and climate adaptation programs.

Such programs must advance with built-in flexibility to respond to evolving priorities.

Conclusion 3: When a community increases its capacity to absorb the effects of hazards and minimizes its recovery needs, disaster effects are less likely to compound.

Targeted, community-guided investments to increase the resilience of essential services and infrastructure are important to achieve this objective. While it is preferable to invest before a disaster occurs, for those communities able to access such investments, the influx of recovery funding after a disaster can provide an opportunity to mitigate future losses and the potential for compounding disasters. As discussed in Chapter 5, many information-gathering session panelists highlighted the importance of flexible pandemic relief funds as an invaluable asset to help their communities recover. The literature on social capital discussed in Chapter 2 and the comments made by information-gathering session panelists both emphasize the critical role of social capital and cohesion in building resilience and catalyzing recovery from disaster events. Investments into building physical capacity to withstand hazards are essential, but the strengthening of adaptive capacity, including formal and informal relationships among community stakeholders, is equally foundational to disaster resilience.

The New Scale and Temporal Scope of Compounding Disasters

Conclusion 4: Perception and understanding of risk are commonly grounded in past experience, leading to complacency in preparation and mitigation.

As discussed in Chapter 3, cognitive biases such as recency bias and normalcy bias can lead to inaccurate conceptualizations of past events and can hamper risk communication, mitigation, and planning for future events that extend beyond what has been experienced or is perceived to be the benchmark extreme. These biases are similarly reflected in emergency management protocols, land-use planning and plans, zoning regulations, public utility design, and building codes, which are often grounded in historical precedent or probabilistic hazard descriptions derived from historical data. Given a changing climate, this hindcasted vantage is unlikely to be representative of future hazard risks. Overcoming these biases requires new strategies.

Conclusion 5: Effective disaster recovery requires an "epoch" rather than "event" view that more fully captures the prolonged effects of compounding disasters and reflects the experienced reality of the community.

In the information-gathering sessions summarized in Chapter 3, panelists continually reported that the damage from and ramifications of discrete disaster events were impossible to disentangle when the disasters occurred in such rapid succession and their recovery time lines overlapped. This was particularly true in socially vulnerable communities that were inadequately resourced before the events occurred. An event-based perspective on disaster management is inherently narrow, reactive, and artificially time constrained. The event-driven view focuses on the symptoms rather than the root causes of disaster losses. Shifting to an epoch view better frames the breadth of lived experiences with disasters and accounts for the potential for compounding losses, driving the broad disaster response and recovery enterprise to more comprehensive and effective pathways forward.

Conclusion 6: Risk assessment and communication for extreme weather-climate and multiple/sequential events is inadequate for both current and future conditions.

The severity of many of the disaster impacts in 2020–2021 exceeded expectations. With a changing climate, the potential for complex events only increases. As described in Chapter 3, many information-gathering session panelists brought to light a variety of disaster communication challenges that were exacerbated by the compounded nature of events, including inconsistent technology and broadband access, accommodating non-English-literate populations, lack of public trust in government, and the spread of misinformation. Better preparation for future compounding events requires incorporation of compound event risk assessments, multisector collaboration, and improved risk communication. The multiple levels of a participatory information-sharing approach will increase understanding of the unique needs of socially vulnerable groups, enhance transparency, and reduce the potential for misinformation.

Bolstering Adaptive Capacity

Conclusion 7: Health care and public health systems will require increased adaptive capacity and staffing to respond to diverse challenges posed by compounding disasters.

Compounding disasters notwithstanding, health systems in the GOM region are already vulnerable and struggle to meet the needs of GOM populations, as discussed in Chapter 2. Recent literature and information-gathering session panelists made clear the increased strain on health systems caused by the co-occurrence of natural hazards and the COVID-19 pandemic. The nature of the hazards encountered influences the effects on population health, health systems, and public health services. Even as extreme weather-climate events damage facilities and disrupt access to health care, emerging disease outbreaks require physically intact facilities and enough front-line professionals staffing them so that services can be rapidly shifted to communicable disease care, and local public health capacity can be maintained.

Conclusion 8: Pervasive mental health issues from compounding disasters undermine the adaptive capacity of communities to withstand and effectively recover from disruptive events.

Information-gathering session panelists spoke extensively about the negative mental health effects of compounding disasters on survivors, especially for specific subpopulations like first responders and volunteers. The impacts of disasters on behavioral health are pervasive, extend across a spectrum of severity, and may continue for a prolonged time. Chapter 2 details some of these specific impacts and explores the links between disaster exposure and adverse psychological and behavioral health outcomes. Mental health needs are magnified for those who experience more intense exposure and vulnerability to hazards. Particular attention is necessary to the mental health needs of those who are disproportionately affected psychologically, including children; the elderly; medically high-risk patients, including those with severe mental illness; and the frontline professionals including first responders, public health professionals, and volunteers tasked with disaster response and recovery responsibilities.

Conclusion 9: The heavy dependence on community-based organizations can strain these individuals and groups beyond the point of effectiveness in the face of compounding disasters.

In the aftermath of a disaster, many immediate needs are met by neighbors helping neighbors and community-based organizations. As discussed in Chapter 4, information-gathering session panelists, particularly those representing CBOs, consistently reported feeling burned out and overwhelmed from spending long hours providing physically and/or emotionally taxing labor for little compensation. While these volunteers and organizations are essential to disaster response and recovery, they may suffer from the compounded disaster impacts themselves and be unable to assist when needed most. Overreliance on this often uncompensated or underpaid workforce risks the depletion of critical human resource capacity for effective long-term recovery from successive disasters.

New Approaches to Contending with Compounding Disasters

Conclusion 10: While a powerful tool for delivering services in times of crisis, technology is not a universal substitute for interpersonal communication and in-person disaster recovery assistance.

COVID-19 restrictions forced rapid innovations, creating more efficient ways to share critical data digitally and greater agility in delivering services virtually—adaptive shifts that were invaluable to service continuity when storms disrupted physical operations. However, as discussed in Chapter 4, processing federal disaster recovery assistance requests and insurance claims virtually caused errors and inefficiencies for survivors. Vulnerable populations, notably the elderly and low–socioeconomic status individuals, are often least equipped to navigate complex and continually evolving online systems. For these populations, the reduction or elimination of in-person assistance prolonged disaster recovery efforts and increased risk for future disruptive events.

Conclusion 11: Safe, sanitary, and secure housing is a fundamental determinant of disaster resilience and recovery. and can thus be viewed as core community infrastructure.

As established in Chapter 2, access to stable, long-term housing is critical for individual and collective well-being, and displacement from secure housing has deleterious effects on security, mental health, social connectedness, and well-being broadly. Since the most vulnerable housing (older/deferred-maintenance properties) is often occupied by the most vulnerable households, this segment of the inventory is especially susceptible to compounding disasters, even as the result of lower-intensity weather-climate events. Areas experiencing economic downturns or with aging housing inventories require particular attention to housing quality as part of ongoing efforts to address the broader affordable housing crisis that existed well before 2020.

Conclusion 12: Effective community-guided risk reduction for compounding disasters requires greater understanding of and planning for the full range of potential disruptive events, along with their cumulative effects.
Chapter 2 outlines the wide range of both hazards to which GOM populations are exposed and the consequences that these hazards can have on human health and well-being. These can be low- or high-probability events, and can span climatic (e.g., rapidly intensifying storms, slow-moving and stalling storms dropping significant rainfall, temperature extremes) and non-climatic (e.g., pandemic, technological) scenarios, including worst-case extremes. Local-level participatory planning processes that routinely engage and collaborate with socially vulnerable communities marginalized by income; education; age; ethnicity/race; representation in policy, governance, and recovery planning; gender; sexual identity; and/or medical risk and who are disproportionately affected by compounding disasters will more effectively guide and prioritize efforts to reduce potential impacts and concurrently build adaptive capacity.

Conclusion 13: Stronger mechanisms are essential to translate lessons recognized from prior experience into lessons learned and implemented.
As discussed in Chapter 5, lessons are often identified after disasters, but they are rarely codified into formal policies and procedures before the next disaster strikes. Interaction between professionals and community members can relay local disaster memories to agency personnel. Ongoing and inclusive education, training, drills, and scenario-based exercises can perpetuate and reinforce lessons learned among all affected communities and decision-makers. Collaborative input of lessons learned into after-action reports can increase knowledge base, participation, and potential for implementation. Efforts at all levels of government to dissolve institutional silos and reinforce improved practices can also sustain lessons learned between disasters.

Conclusion 14: Revisions to disaster planning, response, and recovery policies and procedures need to directly address and eliminate the uneven access to resources that can exacerbate social and economic inequities in the wake of disasters.

Residents in affected areas with the most financial means, social capital, and technological skills are able to more effectively access recovery resources; compounding disasters can leave those without such critical capabilities in a more distressed condition and even minor disruptive events can become disasters. As discussed in Chapter 2, a large portion of GOM residents are in some capacity marginalized, and the current disaster recovery process often exacerbates vulnerabilities rather than addressing these vulnerabilities at their root. Information-gathering session panelists highlighted the need to incorporate equity into the disaster recovery process to better assist marginalized, socially vulnerable community members.

References

Achenbach, J., and D. Keating. 2023. New CDC life expectancy data shows painfully slow rebound from COVID. *The Washington Post*, November 29. https://www.washingtonpost.com/health/2023/11/29/life-expectancy-2022-united-states/ (accessed January 19, 2024).

Achenbach, J., D. Keating, L. McGinley, A. Johnson, and J. Chikwendiu. 2023. Dying early—America's life expectancy crisis: An epidemic of chronic illness is killing us too soon. *The Washington Post*, October 3. https://www.washingtonpost.com/health/interactive/2023/american-life-expectancy-dropping/?itid=lk_interstitial_manual_16 (accessed January 19, 2024).

Adams, R. M., D. P. Eisenman, and D. Glik. 2019. Community advantage and individual self-efficacy promote disaster preparedness: A multilevel model among persons with disabilities. *International Journal of Environmental Research and Public Health* 16(15):2779.

Adams-Fuller, T. 2023. Extreme heat is deadlier than hurricanes, floods and tornadoes combined. *Scientific American*, July 1. https://www.scientificamerican.com/article/extreme-heat-is-deadlier-than-hurricanes-floods-and-tornadoes-combined/ (accessed January 22, 2024).

Adger, W. N. 2000. Social and ecological resilience: Are they related? *Progress in Human Geography* 24(3):347–364.

Adger, W. N., J. Barnett, F. S. Chapin III, and H. Ellemor. 2011. This must be the place: Underrepresentation of identity and meaning in climate change decision-making. *Global Environmental Politics* 11(2):1–25.

Adger, W. N., T. P. Hughes, C. Folke, S. R. Carpenter, and J. Rockstrom. 2005. Social-ecological resilience to coastal disasters. *Science* 309(5737):1036–1039.

Ahmadiani, M., S. Ferreira, and C. E. Landry. 2019. Flood insurance and risk reduction: Market penetration, coverage, and mitigation in coastal North Carolina. *Southern Economic Journal* 85(4):1058–1082.

AHRQ (Agency for Healthcare Research and Quality). 2020. Impact of hurricanes on injury-related emergency department visits, 2005-2016. https://hcup-us.ahrq.gov/reports/statbriefs/sb267-Hurricanes-Injuries-ED-Visits-2005-2016.jsp#:~:text=The%20rate%20of%20injury%2Drelated,increase%20for%20adults%20aged%2065 (accessed January 17, 2024).

Airriess, C. A., W. Li, K. J. Leong, A. C.-C. Chen, and V. M. Keith. 2008. Church-based social capital, networks and geographical scale: Katrina evacuation, relocation, and recovery in a New Orleans Vietnamese American community. *Geoforum* 39(3):1333–1346.

Alabama Department of Economic and Community Affairs. 2022. *2022 DRGR Public Action Plan.* https://adeca.alabama.gov/wp-content/uploads/Alabama-DRGR-Public-Action-Plan-Approved-1.12.23.pdf (accessed November 16, 2023).

Aldhous, P., S. M. Lee, and Z. Hirji. 2021. The Texas winter storm and power outages killed hundreds more people than the state says. *BuzzFeedNews*, May 26. https://www.buzzfeednews.com/article/peteraldhous/texas-winter-storm-power-outage-death-toll (accessed December 1, 2023).

Aldrich, D. P. 2012. *Building resilience: Social capital in post-disaster recovery.* Chicago: University of Chicago Press.

Aldrich, D. P., and M. Meyer. 2015. Social capital and community resilience. *American Behavioral Scientist* 59(2):254–269.

Alemazkoor, N., B. Rachunok, D. R. Chavas, A. Staid, A. Louhghalam, R. Nateghi, and M. Tootkaboni. 2020. Hurricane-induced power outage risk under climate change is primarily driven by the uncertainty in projections of future hurricane frequency. *Scientific Reports* 10:15270.

Allen, B. L. 2007. Environmental justice and expert knowledge in the wake of a disaster. *Social Studies of Science* 37(1):103–110.

Alnajar, A., O. H. Frazier, A. Elgalad, P. A. Smith, and J. M. Shultz. 2021. Preparing end-stage heart failure patients and care providers in the era of climate change-driven hurricanes. *Journal of Cardiac Surgery* 36(10):3491–3493.

Alter, L. 2021. So what ever happened to Katrina cottages? *Treehugger*, February 7. https://www.treehugger.com/so-what-ever-happened-katrina-cottages-4857453 (accessed March 11, 2024).

Andersen, T., and J. M. Shepherd. 2013. A global spatio-temporal analysis of inland tropical cyclone maintenance or intensification. *International Journal of Climatology* 34:391–402.

Arnold, C. A. 2011. Legal castles in the sand: The evolution of property law, culture, and ecology in coastal lands. *Syracuse Law Review* 61:213–260.

ASCE (American Society of Civil Engineers). 2017. *Report card for Louisiana infrastructure.* https://infrastructurereportcard.org/state-item/louisiana/ (accessed January 22, 2024).

ASCE. 2020. *Report card for Mississippi's infrastructure.* https://infrastructurereportcard.org/state-item/mississippi/ (accessed January 22, 2024).

ASCE. 2021a. *Florida infrastructure report card.* https://infrastructurereportcard.org/state-item/florida/ (accessed May 19, 2024).

ASCE. 2021b. *Texas infrastructure report card.* https://infrastructurereportcard.org/state-item/texas/ (accessed January 22, 2024).

ASCE. 2022a. *Minimum design loads and associated criteria for buildings and other structures, ASCE/SEI 7-22.* https://ascelibrary.org/doi/10.1061/9780784415788 (accessed June 15, 2023).

ASCE. 2022b. *Report card for Alabama's infrastructure.* https://infrastructurereportcard.org/state-item/alabama/ (accessed January 22, 2024).

Ash, K. D., M. J. Egnoto, S. M. Strader, W. S. Ashley, D. B. Roueche, K. E. Klockow-McClain, D. Caplen, and M. Dickerson. 2020. Structural forces: Perception and vulnerability factors for tornado sheltering within mobile and manufactured housing in Alabama and Mississippi. *Weather, Climate, and Society* 12:453–472.

Ashley, S. T., and W. S. Ashley. 2008. Flood fatalities in the United States. *Journal of Applied Meteorology and Climatology* 47(3):805–818.

Ashley, W. S. 2007. Spatial and temporal analysis of tornado fatalities in the United States: 1880 2005. *Weather and Forecasting* 22:1214–1228.

Ashley, W. S., and S. M. Strader. 2016. Recipe for disaster: How the dynamic ingredients of risk and exposure are changing the tornado disaster landscape. *Bulletin of the American Meteorological Society* 97:767–786.

Ashley, W. S., S. Strader, T. Rosencrants, and A. J. Krmenec. 2014. Spatiotemporal changes in tornado hazard exposure: The case of the expanding bull's-eye effect in Chicago, Illinois. *Weather, Climate, and Society* 6(2):175–193.

Atoba, K., G. Newman, and G. Sansom. 2023. Multi-hazard property buyouts: Making a case for the acquisition of flood and contaminant-prone residential properties in Galena Park, TX. *Climate Risk Management* 41:100529.

Augustine, L. A. 2024. Ten years into the Gulf Research Program. *Issues in Science and Technology* 40(3): 75–77.

Aune, K. T., D. Gesch, and G. S. Smith. 2020. A spatial analysis of climate gentrification in Orleans Parish, Louisiana post-Hurricane Katrina. *Environmental Research* 185:109384.

Balaguru, K., G. R. Foltz, and L. R. Leung. 2018. Increasing magnitude of hurricane rapid intensification in the central and eastern tropical Atlantic. *Geophysical Research Letters* 45(9):4238–4247.

Balaguru, K., W. Xu, C. C. Chang, L. R. Leung, D. R. Judi, S. M. Hagos, M. F. Wehner, J. P. Kossin, and M. Ting. 2023. Increased US coastal hurricane risk under climate change. *Science Advances* 9(14):eadf0259.

Balbus, J. M., and C. Malina. 2009. Identifying vulnerable subpopulations for climate change health effects in the United States. *Journal of Occupational and Environmental Medicine* 51(1):33–37.

Bankoff, G. 2004. The historical geography of disaster: "Vulnerability" and "local knowledge" in Western discourse. In *Mapping vulnerability: Disasters, development and people,* edited by G. Bankoff, G. Freks, and D. Hilhorst. United Kingdom: Earthscan. Pp. 25–37.

Barlow, M., and S. Camargo. 2022. *State of the planet.* Columbia Climate School. https://news.climate.columbia.edu/2022/10/03/heres-what-we-know-about-how-climate-change-fuels-hurricanes/ (accessed October 3, 2023).

Barry, M. 2023. Triple-I: Louisiana's insurance crisis grew after 2020–21 hurricanes. Press release, March 28, 2023, Insurance Information Institute. https://www.iii.org/press-release/triple-i-louisianas-insurance-crisis-grew-after-2020-21-hurricanes-032823 (accessed October 9, 2023).

Baurick, T. 2019. *Welcome to "cancer alley," where toxic air is about to get worse.* ProPublica. https://www.propublica.org/article/welcome-to-cancer-alley-where-toxic-air-is-about-to-get-worse (accessed May 29, 2024).

Baussan, D. 2015. *When you can't go home: The Gulf Coast 10 years after Katrina.* Center for American Progress, August 18. https://cdn.americanprogress.org/wp-content/uploads/2015/08/18102058/ClimateDisplacement.pdf (accessed June 26, 2023).

Benedict, S. 2022. *Life in Louisiana's forever storm.* The Weather Channel, September 8. https://weather.com/photos/news/2022-08-26-lake-charles-louisiana-hurricane-laura-storm (accessed October 19, 2023).

Bergstrand, K., and B. Mayer. 2020. The community helped me: Community cohesion and environmental concerns in personal assessments of post-disaster recovery. *Society and Natural Resources* 33(3):386–405.

Bernier, C., J. R. Elliott, J. E. Padgett, F. Kellerman, and P. B. Bedient. 2017. Evolution of social vulnerability and risks of chemical spills during storm surge along the Houston Ship Channel. *Natural Hazards Review* 18(4):04017013.

Best, K. B., Z. Jouzi, M. S. Islam, T. Kirby, R. Nixon, A. Hossan, and R. A. Nyiawung. 2023. Typologies of multiple vulnerabilities and climate gentrification across the East Coast of the United States. *Urban Climate* 48:101430.

Bhatia, K., A. Baker, W. Yang, G. Vecchi, T. Knutson, H. Murakami, J. Kossin, K. Hodges, K. Dixon, B. Bronselaer, and C. Whitlock. 2022. A potential explanation for the global increase in tropical cyclone rapid intensification. *Nature Communications* 13:6626.

Bidadian, B., M. P. Strager, P. Butler, and H. Ghadimi. 2024. Flood risk impacts from an unlikely source: Redlining efforts of the 1930s in Houston, Texas. *Environmental Justice* 10.1089/env.2023.0027.

Bikomeye, J. C., S. Namin, C. Anyanwu, C. S. Rublee, J. Ferschinger, K. Leinbach, P. Lindquist, A. Hoppe, L. Hoffman, J. Hegarty, and D. Sperber. 2021. Resilience and equity in a time of crises: Investing in public urban greenspace is now more essential than ever in the US and beyond. *International Journal of Environmental Research and Public Health* 18(16):8420.

Billings, S. B., E. Gallagher, and L. Ricketts. 2019. Let the rich be flooded: The unequal impact of Hurricane Harvey on household debt. *SSRN* 3396611.

Birkmann, J., and N. Fernando. 2008. Measuring revealed and emergent vulnerabilities of coastal communities to tsunamis in Sri Lanka. *Disasters* 32(1):82–104.

Blum, M. F., Y. Feng, G. B. Anderson, D. L. Segev, M. McAdams-DeMarco, and M. E. Grams. 2022. Hurricanes and mortality among patients receiving dialysis. *Journal of the American Society of Nephrology* 33(9):1757–1766.

Bourdieu, P. 1985. The forms of capital. In *Handbook of theory and research for the sociology of education,* edited by J. G. Richardson. Westport, CT: Greenwood. Pp. 241–258.

Bowdler, J., and B. Harris. 2022. *Racial inequality in the United States.* U.S. Department of Treasury, July 21. https://home.treasury.gov/news/featured-stories/racial-inequality-in-the-united-states (accessed October 25, 2023).

Brand, A. L., and K. Seidman. 2008. *Assessing post-Katrina recovery in New Orleans: Recommendations for equitable rebuilding.* Department of Urban Studies and Planning, Massachusetts Institute of Technology. https://www.colab.mit.edu/resources-1/2008/1/7/assessing-post-katrina-recovery-in-new-orleans-recommendations-for-equitable-rebuilding (accessed October 25, 2023).

Brennan, M., T. Srini, J. Steil, M. Mazereeuw, and L. Ovalles. 2021. A perfect storm? Disasters and evictions. *Housing Policy Debate* 32(1):52–83.

Broderick, T. 2023. How fires, floods and hurricanes create deadly pockets of information isolation. *Scientific American*, September 11. https://www.scientificamerican.com/article/how-fires-floods-and-hurricanes-create-deadly-pockets-of-information-isolation/ (accessed January 24, 2024).

Bullard, R. D. 2005. *The quest for environmental justice: Human rights and the politics of pollution* (First edition). Sierra Club Books.

Bullard, R. D., and B. Wright. 2012. *The wrong complexion for protection: How the government response to disaster endangers African American communities.* New York: NYU Press.

Burby, R. J. 2006. Hurricane Katrina and the paradoxes of government disaster policy: Bringing about wise governmental decisions for hazardous areas. *Annals of the American Academy of Political and Social Science* 604(1):171–191.

Burger, M., J. Wentz, and R. Horton. 2021. The law and science of climate change attribution. *Columbia Journal of Environmental Law* 45(1):107916.

Burley, D. M. 2010. *Losing ground: Identity and land loss in coastal Louisiana.* Jackson: University Press of Mississippi.

Burton, I. 2010. Forensic disaster investigations in depth: A new case study model. *Environment* 52(5):36–41.

Busby, J. W., K. Baker, M. D. Bazilian, A. Q. Gilbert, E. Grubert, V. Rai, J. D. Rhodes, S. Shidore, C. A. Smith, and M. E. Webber. 2021. Cascading risks: Understanding the 2021 winter blackout in Texas. *Energy Research & Social Science* 77:102106.

Byrwa-Hill, B. M., T. L. Morphew, A. A. Presto, J. P. Fabisiak, and S. E. Wenzel. 2023. Living in environmental justice areas worsens asthma severity and control: Differential interactions with disease duration, age at onset, and pollution. *Journal of Allergy and Clinical Immunology* 152(5):1321–1329.

Cappucci, M. 2020. Miami is shattering heat records during a wildly-warm start to 2020, even by Florida standards. *The Washington Post*, April 15. https://www.washingtonpost.com/weather/2020/04/15/record-miami-florida-heat/ (accessed January 25, 2024).

Casey, E. S. 1997. *The fate of place: A philosophical history.* Oakland: University of California Press.

Castellanos, S., J. Potts, H. Tiedmann, S. Alverson, Y. R. Glazer, A. Robison, S. Russo, D. Harmon, B. Ken-Opurum, M. Weisz, and F. Acuna. 2023. A synthesis and review of exacerbated inequities from the February 2021 winter storm (Uri) in Texas and the risks moving forward. *Progress in Energy* 5(1):012003.

CDC (Centers for Disease Control and Prevention). 2022. *Life expectancy at birth by state.* https://www.cdc.gov/nchs/pressroom/sosmap/life_expectancy/life_expectancy.htm (accessed February 14, 2024).

CDC. 2023a. *Infant mortality rates by state.* https://www.cdc.gov/nchs/pressroom/sosmap/infant_mortality_rates/infant_mortality.htm (accessed February 14, 2024).

CDC. 2023b. *Stats of the states: Causes of death.* https://www.cdc.gov/nchs/pressroom/stats_of_the_states.htm (accessed February 14, 2024).

CDC. 2023c. *COVID-19 mortality by state.* https://www.cdc.gov/nchs/pressroom/sosmap/covidmortality_final/COVID19.html (acccessed January 30, 2024).

CDC. 2023d. *COVID-19 vaccinations in the United States, jurisdiction.* https://data.cdc.gov/Vaccinations/COVIDvaccinations-in-the-United-States-Jurisdi/unsk-b7fc/data_preview (accessed February 14, 2024).

CDC. 2024. *CDC/ATSDR Social vulnerability index: Overview.* https://www.atsdr.cdc.gov/placeandhealth/svi/index.html (accessed May 22, 2024).

CDP (Center for Disaster Philanthropy). 2023. *Women and girls in disasters.* https://disaster-philanthropy.org/resources/women-and-girls-in-disasters/ (accessed January 22, 2024).

Cerdá, M., P. M. Bordelois, S. Galea, F. Norris, M. Tracy, and K. C. Koenen. 2013. The course of posttraumatic stress symptoms and functional impairment following a disaster: What is the lasting influence of acute versus ongoing traumatic events and stressors? *Social Psychiatry and Psychiatric Epidemiology* 48(3):385–395.

Chakraborty, J. 2021. Convergence of COVID-19 and chronic air pollution risks: Racial/ethnic and socioeconomic inequities in the U.S. *Environmental Research* 193:110586.

Chakraborty, J., T. W. Collins, and S. E. Grineski. 2019. Exploring the environmental justice implications of Hurricane Harvey flooding in Greater Houston, Texas. *American Journal of Public Health* 109(2):244–250.

Chakraborty, J., S. E. Grineski, and T. W. Collins. 2019. Hurricane Harvey and people with disabilities: Disproportionate exposure to flooding in Houston, Texas. *Social Science & Medicine* 226:176–181.

Chakraborty, J., T. W. Collins, and S. E. Grineski. 2023. Disability and subsidized housing residency: The adverse impacts of Winter Storm Uri in metropolitan Texas. *Disability and Health Journal* 16(2):101403.

Chakraborty, J., S. E. Grineski, T. W. Collins, and J. J. Aun. 2024. Disparities in adverse impacts of the COVID-19 pandemic by disability status in metropolitan Texas. *Journal of Public Health* 46(1):e60–e64.

Chakraborty, J., A. McAfee, T. W. Collins, and S. E. Grineski. 2021. Exposure to Hurricane Harvey flooding for subsidized housing residents in Harris County, Texas. *Natural Hazards* 106(3):2185–2205.

Cherry, K. E., M. R. Calamia, T. Birch, and A. Moles. 2021. Assessing mental health after a disaster: Flood exposure, recovery stressors, and prior flood experience. In *The Intersection of trauma and disaster behavioral health*, edited by K. E. Cherry and A. Gibson. Springer Cham. Pp. 271–283.

Cindass, R., T. S. Cancio, J. M. Cancio, K. A. Prukowski, S. E. Park, S. K. Shingleton, C. M. Yugawa, and L. C. Cancio. 2023. Management of multiple frostbite casualties at a burn center: San Antonio, Texas, 12–20 February 2021. *Burns* 49(8):1990–1996.

City of Houston. 2022. *Draft action plan for disaster recovery: 2021 winter storm.* https://houstontx.gov/housing/dr2021/action-plan/2021_Winter_Storm_Draft_Action_Plan-080522.pdf (accessed September 10, 2023).

Clark-Ginsberg, A., L. C. Easton-Calabria, S. S. Patel, J. Balagna, and L. A. Payne. 2021. When disaster management agencies create disaster risk: A case study of the US's Federal Emergency Management Agency. *Disaster Prevention and Management* 30(4-5):447–461.

Clay, L. A., and D. M. Abramson. 2021. Bowling together: Community social institutions protective against poor child mental health. *Environmental Justice* 14(3):206–215.

Climate Central. 2024. *Weather-related power outages rising.* https://www.climatecentral.org/climate-matters/weather-related-power-outages-rising (accessed June 17, 2024).

CMS (Centers for Medicare & Medicaid Services). 2018. *Chronic conditions prevalence, state/county 2018.* https://cms-oeda.maps.arcgis.com/apps/MapSeries/index.html?appid=062934f815eb412182b3d324054ea6f0 (accessed February 15, 2024).

CMS. n.d. *Accounting for federal COVID expenditures in the National Health Expenditure Accounts.* https://www.cms.gov/files/document/accounting-federal-covid-expenditures-national-health-expenditure-accounts.pdf (accessed July 11, 2024).

Cohen, G. H., R. Wang, L. Sampson, S. R. Lowe, C. K. Ettman, S. M. Abdalla, G. A. Wellenius, H. Cabral, K. Ruggiero, and S. Galea. 2023. Depression and PTSD among Houston residents who experienced Hurricane Harvey and COVID-19: Implications for urban areas affected by multiple disasters. *Journal of Urban Health* 100(4):860–869.

Cohen, J., L. Agel, and I. White. 2021. Linking Arctic variability and change with extreme winter weather in the United States. *Science* 373(6559):1116–1121.

Collins, J., A. Polen, K. McSweeney, D. Colón-Burgos, and I. Jernigan. 2021. Hurricane risk perceptions and evacuation decision-making in the age of COVID-19. *Bulletin of the American Meteorological Society* 102(4):E836–E848.

Collins, T. W., S. E. Grineski, J. Chakraborty, and A. B. Flores. 2019. Environmental injustice and Hurricane Harvey: A household-level study of socially disparate flood exposures in Greater Houston, Texas, USA. *Environmental Research* 179:108772.

Colten, C. E. 2005. *Unnatural metropolis: Wresting New Orleans from nature.* Baton Rouge: LSU Press.

Colten, C. E. 2009. *Perilous place, powerful storms: Hurricane protection in coastal Louisiana.* Baton Rouge: LSU Press.

Colten, C. E. 2012. Forgetting the unforgettable: Losing resilience in New Orleans. In *American environments: Climate—cultures—catastrophe,* edited by C. Mauch and M. Sylvia. Heidelberg: Universitätsverlag. Pp. 159–176.

Colten, C. E. 2015. Reshaping the riparian: Human mobility and fixed infrastructure. *Confins* 23:10082.

Colten, C. E. 2020. Eroding memories and erecting risk on the Amite River. *Open Rivers: Rethinking Water, Place & Community* 16:6–23.

Colten, C. E. 2021. As inland becomes coastal: Shifting equity and flood risk in the Amite River Basin (USA). *Global Environment* 14(3):475–504.

Colten, C. E., and A. Giancarlo. 2011. Losing resilience on the Gulf Coast: Hurricanes and social memory. *Environment: Science and Policy for Sustainable Development* 53(4):6–19.

Colten, C. E., A. A. Grismore, and J. R. Simms. 2015. Oil spills and community resilience: Uneven impacts and protection in historical perspective. *Geographical Review* 105(4):391–407.

Colten, C. E., J. Hay, and A. Giancarlo. 2012. Community resilience and oil spills in coastal Louisiana. *Ecology and Society* 17(3):5.

Commonwealth Fund. 2023. *Scorecard on state health system performance: Americans' health declines and access to reproductive care shrinks, but states have options.* https://doi.org/10.26099/fcas-cd24 (accessed October 4, 2023).

Copernicus. 2024. 2023 is the hottest year on record, with global temperatures close to the 1.5 °C limit. Press release, January 9, 2024. https://climate.copernicus.eu/copernicus-2023-hottest-year-record (accessed January 11, 2024).

Craig, R. K. 2019. Coastal adaptation, government-subsidized insurance, and perverse incentives to stay. *Climatic Change* 152(2):215–226.

CRS (Congressional Research Service). 2023. *Is that climate change? The science of extreme event attribution.* https://crsreports.congress.gov/product/pdf/R/R47583 (accessed October 25, 2023).

CSB (U.S. Chemical Safety Board). 2017. Investigation report. Organic peroxide decomposition, release, and fire at Arkema Crosby following Hurricane Harvey flooding. Fact Sheet. https://www.arkema.com/files/live/sites/arkema_usa/files/downloads/social-responsibility/key-facts-csb-100-year-500-year-floods.pdf (accessed June 13, 2024).

Cutter, S. L. 2018. Compound, cascading, or complex disasters: What's in a name? *Environment: Science and Policy for Sustainable Development* 60(6):16–25.

Cutter, S. L., L. Barnes, M. Berry, C. Burton, E. Evans, E. Tate, and J. Webb. 2008. A place-based model for understanding community resilience to natural disasters. *Global Environmental Change* 18(4):598–606.

Cutter, S. L., B. L. Boruff, and W. L. Shirley. 2003. Social vulnerability to environmental hazards. *Social Science Quarterly* 84(2):242–261.

Cutter, S. L., C. T. Emrich, J. T. Mitchell, W. W. Piegorsch, M. M. Smith, and L. Weber. 2014. *Hurricane Katrina and the forgotten coast of Mississippi.* Cambridge: Cambridge University Press.

Dangendorf, S., N. Hendricks, Q. Sun, J. Klinck, T. Frederikse, F. M. Calafat, T. Wahl, and T. E. Törnqvist. 2023. Acceleration of U.S. Southeast and Gulf Coast sea-level rise amplified by internal climate variability. *Nature Communications* 14:1935.

Davidson, J. R., and A. C. McFarlane. 2006. The extent and impact of mental health problems after disaster. *Journal of Clinical Psychiatry* 67(Suppl 2):9–14.

de Ruiter, M. C., A. Couasnon, M. J. van den Hombert, J. E. Daniell, J. C. Gill, and P. J. Ward. 2020. Why we can no longer ignore consecutive disasters. *Earth's Future* 8:1–9.

Deen, A. 2021. What is heirs' property? A huge contributor to Black land loss you might not have heard of. *Grist*, March 17. https://grist.org/fix/justice/what-is-heirs-property-a-huge-contributor-to-black-land-loss-you-might-not-have-heard-of/ (accessed December 14, 2023).

DeRobertis, J. 2024. In first meeting, new Livingston Parish Council pushes for stricter oversight of developers. *The Advocate*, June 11. https://www.theadvocate.com/baton_rouge/news/new-livingston-parish-council-pushes-to-restrict-development/article_84ddcea6-b0e1-11ee-8be0-37b3db821e84.html (accessed February 2, 2024).

Despart, Z. 2022. Did Texas undercount 2021's freeze deaths? COVID and historical patterns cannot explain spike. *Houston Chronicle*, February 18. https://www.houstonchronicle.com/news/houston-texas/houston/article/Did-Texas-undercount-2021-freeze-deaths-COVID8281.php (accessed December 10, 2023).

Di Liberto, T. 2021. *A foot of rain causes flash flood emergency in Louisiana during mid-May 2021.* U.S. Department of Commerce, May 21. https://www.climate.gov/news-features/event-tracker/foot-rain-causes-flash-flood-emergency-louisiana-during-mid-may-2021 (accessed January 24, 2024).

Dixon, P. G., A. E. Mercer, J. Choi, and J. S. Allen. 2011. Tornado risk analysis: Is Dixie Alley an extension of Tornado Alley? *Bulletin of the American Meteorological Society* 92:433–441.

Do, V., H. McBrien, N. M. Flores, A. J. Northrop, J. Schlegelmilch, M. V. Kiang, and J. A. Casey. 2023. Spatiotemporal distribution of power outages with climate events and social vulnerability in the USA. *Nature Communications* 14(1):2470.

Domingue, S. J. 2022. The (in)dispensability of environmental justice communities: A case study of climate adaptation injustices in coastal Louisiana and narratives of resistance. *Environmental Justice* 15(4):271–278.

Domingue, S. J., and C. T. Emrich. 2019. Social vulnerability and procedural equity: Exploring the distribution of disaster aid across counties in the United States. *American Review of Public Administration* 49(8):897–913.

Donaghy, T. Q., N. Healy, C. Y. Jiang, and C. P. Battle. 2023. Fossil fuel racism in the United States: How phasing out coal, oil, and gas can protect communities. *Energy Research & Social Science* 100:103104.

Donahue, A., and R. Tuohy. 2006. Lessons we don't learn: A study of the lessons of disasters, why we repeat them and how we can learn them. *Homeland Security Affairs* 2(2):1–28.

Donald, J. 2021. The economic impact of Winter Storm Uri. *Corridor News,* November 3. https://smcorridornews.com/the-economic-impact-of-winter-storm-uri/ (accessed February 13, 2024).

Doss-Gollin, J., D. J. Farnham, U. Lall, and V. Modi. 2021. How unprecedented was the February 2021 Texas cold snap? *Environmental Research Letters* 16:064056.

Drabek, T. E. 2007. Social problems perspectives, disaster research and emergency management: Intellectual contexts, theoretical extensions, and policy implications. *Annual Meeting of the American Sociological Association* 1–117.

Dreier, H., and A. B. Tran. 2021. "The real damage": Why FEMA is denying disaster aid to Black families that have lived for generations in the deep South. *The Washington Post,* July 11. https://www.washingtonpost.com/nation/2021/07/11/fema-black-owned-property/ (accessed October 25, 2023).

Easterling, D. R., K. E. Kunkel, J. R. Arnold, T. Knutson, A. N. LeGrande, L. R. Leung, R. S. Vose, D. E. Waliser, and M. F. Wehner. 2017. Precipitation change in the United States. In *Climate science special report: Fourth national climate assessment,* vol. 1, edited by D. J. Wuebbles, D. W. Fahey, K. A. Hibbard, D. J. Dokken, B. C. Stewart, and T. K. Maycock. Washington, DC: U.S. Global Change Research Program. Pp. 207–230.

Ebi, K. L., J. Vanos, J. W. Baldwin, J. E. Bell, D. M. Hondula, N. A. Errett, K. Hayes, C. E. Reid, S. Saha, J. Spector, and P. Berry. 2021. Extreme weather and climate change: Population health and health system implications. *Annual Review of Public Health* 42:293–315.

EDF (Environmental Defense Fund). 2023. U.S. Climate Vulnerability Index shows where action, resources are most urgently needed. Press release, October 2. https://www.edf.org/cvi-news (accessed January 8, 2024).

Elder, K., S. Xirasagar, N. Miller, S. A. Bowen, S. Glover, and C. Piper. 2007. African Americans' decision not to evacuate New Orleans before Hurricane Katrina: A qualitative study. *American Journal of Public Health* 97(Suppl 1):S124–S129. [Errata: *American Journal of Public Health* 97(12):2122.]

Ellen, I. G., J. P. Steil, and J. De La Roca. 2016. The significance of segregation in the 21st century. *City and Community* 15(1):8–13.

Elliott, J. R., T. J. Haney, and P. Sams-Abiodun. 2010. Limits to social capital: Comparing network assistance in two New Orleans neighborhoods devastated by Hurricane Katrina. *Sociological Quarterly* 51(4):624–648.

Emrich, C. T., S. K. Aksha, and Y. Zhou. 2022. Assessing distributive inequities in FEMA's disaster recovery assistance fund allocation. *International Journal of Disaster Risk Reduction* 74:102855.

Erikson, K. 1976. *Everything in its path.* New York: Simon and Schuster.

Ernst, K. C., A. R. Crimmins, S. Anenberg, M. H. Hayden, B. O. Hoppe, L. J. Mickley, D. E. Peck, H. J. Tanana, and J. J. West. 2023. Focus on COVID-19 and climate change. In *Fifth national climate assessment,* edited by A. R. Crimmins, C. W. Avery, D. R. Easterling, K. E. Kunkel, B. C. Stewart, and T. K. Maycock. Washington, DC: U.S. Global Change Research Program.

Errett, N. A., C. Hartwell, J. M. Randazza, A. Nori-Sarma, K. R. Weinberger, K. R. Spangler, Y. Sun, Q. H. Adams, G. A. Wellenius, and J. J. Hess. 2023. Survey of extreme heat public health preparedness plans and response activities in the most populous jurisdictions in the United States. *BMC Public Health* 23(1):811.

Espinel, Z., S. Galea, J. P. Kossin, C. Caban-Aleman, and J. M. Shultz. 2019. Climate-driven Atlantic hurricanes pose rising threats for psychopathology. *Lancet Psychiatry* 6(9):721–723.

Espinel, Z., J. P. Kossin, S. Galea, A. S. Richardson, and J. M. Shultz. 2019. Forecast: Increasing mental health consequences from Atlantic hurricanes throughout the 21st century. *Psychiatric Services* 70(12):1165–1167.

Espinel, Z., L. M. Nogueira, H. A. Gay, J. M. Bryant, W. Hamilton, E. J. Trapido, J. M. Shepherd, S. Galea, and J. M. Shultz. 2022. Climate-driven Atlantic hurricanes create complex challenges for cancer care. *Lancet Oncology* 23(12):1497–1498.

Espinel, Z., J. M. Shultz, V. P. Aubry, O. M. Abraham, Q. Fan, T. E. Crane, L. Sahar, L. M. Nogueira. 2023. Protecting vulnerable patient populations from climate hazards: The role of the nation's cancer centers. *Journal of the National Cancer Institute* 115(11):1252–1261.

Faber, J. W. 2015. Superstorm Sandy and the demographics of flood risk in New York City. *Human Ecology* 43(3):363–378.

Falk, G., P. D. Romero, I. A. Micchitta, and W. C. Nyhof. 2020. *Unemployment rates during the COVID-19 pandemic: In brief.* Congressional Research Service, September 30. https://crsreports.congress.gov/product/pdf/R/R46554/1 (accessed on October 22, 2023).

FEMA (Federal Emergency Management Agency). n.d. *Glossary.* https://www.fema.gov/about/glossary/r (accessed February 27, 2024).

FEMA. 2020. *National Advisory Council report to the FEMA administrator: November 2020.* https://www.fema.gov/sites/default/files/documents/fema_nac-report_11-2020.pdf (accessed October 25, 2023).

FEMA. 2022. FEMA provides multiple ways to prove home ownership. Press Release no. NR-024, October 12. https://www.fema.gov/press-release/20221013/fema-provides-multiple-ways-prove-home-ownership (accessed February 6, 2024).

FEMA. 2024. Impacted by two disasters—what you need to know. Press release no. 001, March 22, 2024. https://www.fema.gov/press-release/20240403/impacted-two-disasters-what-you-need-know (accessed June 17, 2024).

Finch, C., C. T. Emrich, and S. L. Cutter. 2010. Disaster disparities and differential recovery in New Orleans. *Population and Environment* 31:179–202.

Finch, M. 2022. Here are the Louisiana insurers that have gone broke or left the state amid deepening crisis. *NOLA*, September 27. https://www.nola.com/news/business/here-are-the-louisiana-insurers-that-have-gone-broke-or-left-the-state-amid-deepening/article_c7f077b4-3e98-11ed-86c9-f7f11037202f.html (accessed October 9, 2023).

Finucane, M. L., J. Acosta, A. Wicker, and K. Whipkey. 2020. Short-term solutions to a long-term challenge: Rethinking disaster recovery planning to reduce vulnerabilities and inequities. *International Journal of Environmental Research and Public Health* 17(2):482.

First Street Foundation. 2022. *The 6th national risk assessment—Hazardous heat.* https://report.firststreet.org/6th-National-Risk-Assessment-Hazardous-Heat.pdf (accessed December 12, 2023).

Fishback, P., J. Rose, K. A. Snowden, and T. Storrs. 2022. *New evidence on redlining by federal housing programs in the 1930s.* Working paper 29244. National Bureau of Economic Research. https://www.nber.org/system/files/working_papers/w29244/w29244.pdf (accessed February 14, 2024).

Flores, A. B., A. Castor, S. E. Grineski, T. W. Collins, and C. Mullen. 2021. Petrochemical releases disproportionately affected socially vulnerable populations along the Texas Gulf Coast after Hurricane Harvey. *Population and Environment* 42:279–301.

Flores, A. B., T. W. Collins, S. E. Grineski, and J. Chakraborty. 2020a. Disparities in health effects and access to health care among Houston area residents after Hurricane Harvey. *Public Health Reports* 135(4):511–523.

Flores, A. B., T. W. Collins, S. E. Grineski, and J. Chakraborty. 2020b. Social vulnerability to Hurricane Harvey: Unmet needs and adverse event experiences in Greater Houston, Texas. *International Journal of Disaster Risk Reduction* 46:101521.

Flores, A. B., T. W. Collins, S. E. Grineski, A. L. Griego, C. Mullen, S. M. Nadybal, R. Renteria, R. Rubio, Y. Shaker, and S. A. Trego. 2021. Environmental injustice in the disaster cycle: Hurricane Harvey and the Texas Gulf Coast. *Environmental Justice* 14(2):146–158.

Fos, P. J., P. A. Honoré, R. L. Honoré, and K. Patterson. 2021. Health status in fence-line communities: The impact of air pollution. *International Journal of Family Medicine and Primary Care* 2(3):1040.

Fothergill, A., and L. Peek. 2004. Poverty and disasters in the United States: A review of recent sociological findings. *Natural Hazards* 32(1):89–110.

Fowler, H. J., G. Lenderink, A. F. Prein, S. Westra, R. P. Allan, N. Ban, R. Barbero, P. Berg, S Blenkinsop, H. X. Do, S. Guerreiro, J. O. Haerter, E. J. Kendon, E. Lewis, C. Schaer, A. Sharma, G. Villarini, C. Wasko, and X. Zhang. 2021. Anthropogenic intensification of short-duration rainfall extremes. *Nature Reviews Earth & Environment* 2:107–122.

Frazier, T. G., N. Wood, B. Yarnal, and D. H. Bauer. 2010. Influence of potential sea level rise on societal vulnerability to hurricane storm-surge hazards, Sarasota County, Florida. *Applied Geography* 30(4):490–505.

Freeman, A. C., and W. S. Ashley. 2017. Changes in the US hurricane disaster landscape: The relationship between risk and exposure. *Natural Hazards* 88(2):659–682.

Freudenburg, W. R., R. Gramling, S. Laska, and K. T. Erikson. 2011. *Catastrophe in the making: The engineering of Katrina and the disasters of tomorrow.* Washington, DC: Island Press/Center for Resource Economics.

Fullerton, C. S., J. B. McKibben, D. B. Reissman, T. Scharf, K. M. Kowalski-Trakofler, J. M. Shultz, and R. J. Ursano. 2013. Posttraumatic stress disorder, depression, and alcohol and tobacco use in public health workers after the 2004 Florida hurricanes. *Disaster Medicine and Public Health Preparedness* 7(1):89–95.

Fussell, E., S. R. Curran, M. D. Dunbar, M. A. Babb, L. Thompson, and J. Meijer-Irons. 2017. Weather-related hazards and population change: A study of hurricanes and tropical storms in the United States, 1980–2012. *Annals of the American Academy of Political and Social Science* 669(1):146–167.

Gabriel, A. 2023. How to stay safe in your mobile home during a tornado. *Fox Weather*, April 3. https://www.foxweather.com/learn/tornado-proof-mobile-manufactured-home (accessed October 3, 2023).

Galea S., C. R. Brewin, M. Gruber, R. T. Jones, D. W. King, L. A. King, R. J. McNally, R. J. Ursano, M. Petukhova, and R. C. Kessler. 2007. Exposure to hurricane-related stressors and mental illness after Hurricane Katrina. *Archives of General Psychiatry* 64(12):1427–1434.

Gall, M., and C. J. Friedland. 2020. If mitigation saves 6 per every 1 spent, then why are we not investing more? A Louisiana perspective on a national issue. *Natural Hazards Review* 21(1):04019013.

Gao, S., and Y. Wang. 2023. Anticipating older populations' health risk exacerbated by compound disasters based on mortality caused by heart diseases and strokes. *Scientific Reports* 13(1): 16810.

Garfin, D. R., R. R. Thompson, E. A. Holman, G. Wong-Parodi, and R. C. Silver. 2022. Association between repeated exposure to hurricanes and mental health in a representative sample of Florida residents. *JAMA Network Open* 5(6):e2217251.

Gensini, V. A., and H. E. Brooks. 2018. Spatial trends in United States tornado frequency. *npj Climate and Atmospheric Science* 1:38.

Gilbert, J., S. D. Wood, and G. Sharp. 2002. Who owns the land? Agricultural land ownership by race/ethnicity. *Rural America* 17(4):55–62.

Girgin, S., A. Necci, and E. Krausmann. 2019. Dealing with cascading multi-hazard risks in national risk assessment: The case of Natech accidents. *International Journal of Disaster Risk Reduction* 35:101072.

Gissing, A., M. Timms, S. Browning, R. Crompton, and J. McAneney. 2022. Compound natural disasters in Australia: A historical analysis. *Environmental Hazards* 21(2):159–173.

Gitter, A., D. E. Boellstorff, K. D. Mena, D. M. Gholson, K. J. Pieper, C. A. Chavarria, and T. J. Gentry. 2023. Quantitative microbial risk assessment for private wells in flood-impacted areas. *Water* 15(3):469.

Glaser, R., D. Riemann, J. Schönbein, M. Barriendos, R. Brázdil, C. Bertolin, D. Camuffo, M. Deutsch, P. Dobrovolný, A. van Engelen, and S. Enzi. 2010. The variability of European floods since AD 1500. *Climatic Change* 101:235–256.

Gotham, K. F., and M. Greenberg. 2014. *Crisis cities: Disaster and redevelopment of New York and New Orleans.* New York: Oxford University Press.

Gottlieb, L. M., M. S. Pantell, and L. S. Solomon. 2021. The National Academy of Medicine Social Care Framework and COVID-19 care innovations. *Journal of General Internal Medicine* 36(5):1411–1414.

Gradus, J., P. Qin, A. Lincoln, M. Miller, E. Lawler, H. T. Sorensen, and T. Lash. 2010. Posttraumatic stress disorder and completed suicide. *American Journal of Epidemiology* 171(6):721–727.

Graves, G. 2023. Graves and Plaskett introduce bipartisan bill to kick-start recovery for families after natural disasters. Press release, November 13. https://garretgraves.house.gov/news/documentsingle.aspx?DocumentID=2909 (accessed January 24, 2024).

Gray, M. N. 2022. *Arbitrary lines: How zoning broke the American city and how to fix it.* Washington, DC: Island Press.

Griego, A. L., A. B. Flores, T. W. Collins, and S. E. Grineski. 2020. Social vulnerability, disaster assistance, and recovery: A population-based study of Hurricane Harvey in Greater Houston, Texas. *International Journal of Disaster Risk Reduction* 51:101766.

Grineski, S. E., T. W. Collins, and J. Chakraborty. 2022. Cascading disasters and mental health inequities: Winter Storm Uri, COVID-19 and post-traumatic stress in Texas. *Social Science & Medicine* 315:115523.

Grineski, S. E., T. W. Collins, J. Chakraborty, E. Goodwin, J. Aun, and K. D. Ramos. 2023. Social disparities in the duration of power and piped water outages in Texas after Winter Storm Uri. *American Journal of Public Health* 113(1):30–34

Grineski, S. E., A. B. Flores, T. W. Collins, and J. Chakraborty. 2020. Hurricane Harvey and Greater Houston households: Comparing pre-event preparedness with post-event health effects, event exposures, and recovery. *Disasters* 44(2):408–432.

Grineski, S., M. Scott, T. Collins, J. Chakraborty, and K. Ramos. 2023. Anxiety and depression after Winter Storm Uri: Cascading disasters and mental health inequities. *International Journal of Disaster Risk Reduction* 96:103933.

Grove, M., L. Ogden, S. Pickett, C. Boone, G. Buckley, D. H. Locke, C. Lord, and B. Hall. 2018. The legacy effect: Understanding how segregation and environmental injustice unfold over time in Baltimore. *Annals of the American Association of Geographers* 108(2):524–537.

Gruebner, O., S. R. Lowe, M. Tracy, M. Cerdá, S. Joshi, F. H. Norris, and S. Galea. 2016. The geography of mental health and general wellness in Galveston Bay after Hurricane Ike: A spatial epidemiologic study with longitudinal data. *Disaster Medicine and Public Health Preparedness* 10(2):261–273.

Gruebner, O., S. R. Lowe, M. Tracy, S. Joshi, M. Cerdá, F. H. Norris, S. V. Subramanian, and S. Galea. 2016. Mapping concentrations of posttraumatic stress and depression trajectories following Hurricane Ike. *Scientific Reports* 6:32242.

Guidi, J., M. Lucente, N. Sonino, and G. A. Fava. 2021. Allostatic load and its impact on health: A systematic review. *Psychotherapy and Psychosomatics* 90(1):11–27.

Guido, Z., T. Allen, S. Mason, and P. Méndez-Lázaro. 2022. Hurricanes and anomalous heat in the Caribbean. *Geophysical Research Letters* 49(21):101029.

Hahn, M. B., R. Van Wyck, L. Lessard, and R. Fried. 2022. Compounding effects of social vulnerability and recurring natural disasters on mental and physical health. *Disaster Medicine and Public Health Preparedness* 16(3):1013–1021.

Hall, T., and J. P. Kossin. 2019. Hurricane stalling along the North American coast and implications for rainfall. *npj Climate and Atmospheric Science* 2:17.

Hamideh, S., and J. Rongerude. 2018. Social vulnerability and participation in disaster recovery decisions: Public housing in Galveston after Hurricane Ike. *Natural Hazards* 93:1629–1648.

Han, Y., H. Jia, C. Xu, M. Bockarjova, C. van Westen, and L. Lombardo. 2024. Unveiling spatial inequalities: Exploring county-level disaster damages and social vulnerability on public disaster assistance in contiguous US. *Journal of Environmental Management* 351:119690

Hanchey, A., S. Jiva, T. Bayleyegn, and A. Schnall. 2023. Mortality surveillance during Winter Storm Uri, United States—2021. *Disaster Medicine and Public Health Preparedness* 17:e530.

Hanna, J., S. Almasy, and M. Holcombe. 2021. If you live in a state with a low vaccination rate, you're 4 times more likely to be hospitalized and more than 5 times more likely to die. *CNN Health*, August 19. https://www.cnn.com/2021/08/18/health/us-coronavirus-wednesday/index.html (accessed November 23, 2021).

Harris, K., F. Lager, M. K. Jansen, and M. Benzie. 2022. Rising to a new challenge: A protocol for case-study research on transboundary climate risk. *Weather, Climate, and Society* 14:755–768.

Hartwig, R. P., and C. Wilkinson. 2016. *Residual market property plans: From markets of last resort to markets of first choice.* https://www.iii.org/sites/default/files/docs/pdf/residual_markets_wp_051616.pdf (accessed March 11, 2020).

Harvey, C. 2024. Record-breaking ocean heat wave foreshadows a dangerous hurricane season. *Scientific American*, May 13. https://www.scientificamerican.com/article/record-breaking-ocean-heat-wave-foreshadows-a-dangerous-hurricane-season/#:~:text=Extreme%20ocean%20temperatures%20are%20jet,warmest%20parts%20of%20the%20year (accessed June 10, 2024).

Haworth, B. T., S. McKinnon, and C. Eriksen. 2022. Advancing disaster geographies: From marginalisation to inclusion of gender and sexual minorities. *Geography Compass* 16(11):e12664.

Heagele, T., and D. Pacquiao. 2019. Disaster vulnerability of elderly and medically frail populations. *Health Emergency and Disaster Nursing* 6(1):50–61.

Hedquist, A., R. N. Salas, M. Soto, E. J. Orav, and J. F. Figueroa. 2023. Disparities in property insurance relief among socially vulnerable Texas communities after Winter Storm Uri. *Health Affairs Scholar* 1(5):1–4.

Hendricks, M., and S. Van Zandt. 2021. Unequal protection revisited: Planning for environmental justice, hazard vulnerability, and critical infrastructure in communities of color. *Environmental Justice* 14(2):87–97.

Henry, M., N. Spencer, and E. Strobl. 2020. The impact of tropical storms on households: Evidence from panel data on consumption. *Oxford Bulletin of Economics and Statistics* 82(1):1–22.

Herberman Mash, H. B., C. S. Fullerton, K. Kowalski-Trakofler, D. B. Reissman, T. Scharf, J. M. Shultz, and R. J. Ursano. 2013. Florida Department of Health workers' response to 2004 hurricanes: A qualitative analysis. *Disaster Medicine and Public Health Preparedness* 7(2):153–159.

Hersher, R., and R. Kellman. 2021. Why FEMA aid is unavailable to many who need it the most. *Short Wave Science*. National Public Radio, June 29. https://www.npr.org/2021/06/29/1004347023/why-fema-aid-is-unavailable-to-many-who-need-it-the-most (accessed September 20, 2023).

Hess, J. J., J. N. Malilay, and A. J. Parkinson. 2008. Climate change: The importance of place. *American Journal of Preventive Medicine* 35(5):468–478.

Hirschberger, G. 2018. Collective trauma and the social construction of meaning. *Frontiers in Psychology*, 1441.

Hlávka, J., and A. Rose. 2022. *COVID-19's total cost to the U.S. economy will reach $14 trillion by end of 2023. The evidence base.* USC Center for Economic and Social Research. https://healthpolicy.usc.edu/article/covid-19s-total-cost-to-the-economy-in-us-will-reach-14-trillion-by-end-of-2023-new-research/ (accessed July 11, 2024).

Hoffman, J. S., S. G. McNulty, C. Brown, K. D. Dello, P. N. Knox, A. Lascurain, C. Mickalonis, G. T. Mitchum, L. Rivers III, M. Schaefer, G. P. Smith, J. S. Camp, and K. M. Wood. 2023. Ch. 22. Southeast. In *Fifth national climate assessment*, edited by A. R. Crimmins, C. W. Avery, D. R. Easterling, K. E. Kunkel, B. C. Stewart, and T. K. Maycock. Washington, DC: U.S. Global Change Research Program.

Horney, J. A., G. A. Casillas, E. Baker, K. W. Stone, K. R. Kirsch, K. Camargo, T. L. Wade, and T. J. McDonald. 2018. Comparing residential contamination in a Houston environmental justice neighborhood before and after Hurricane Harvey. *PLOS ONE* 13(2):e0192660.

Horowitz, A. 2020. *Katrina: A history, 1915–2015*. Cambridge, MA: Harvard University Press.

Howe, N. 2019. More women die in natural disasters—Why? And what can be done? *Brink News*, April 25. https://www.brinknews.com/gender-and-disasters/?utm_source=BRINK+Asia (accessed September 5, 2023).

Howell, J., and J. R. Elliott. 2019. Damages done: The longitudinal impacts of natural hazards on wealth inequality in the United States. *Social Problems* 66(3):448–467.

HRSA (Health Resources and Services Administration). 2023. *What is a Health Professional Shortage Area (HPSA)?* https://bhw.hrsa.gov/workforce-shortage-areas/shortage-designation#hpsas (accessed July 11, 2024).

Hsu, M., R. Howitt, and F. Miller. 2015. Procedural vulnerability and institutional capacity deficits in post-disaster recovery and reconstruction: Insights from Wutai Rukai experiences of Typhoon Morakot. *Human Organization* 74(4):308–318.

Hu, G., N. Hamovit, K. Croft, and D. Niemeier. 2022. Assessing inequities underlying racial disparities of COVID-19 mortality in Louisiana parishes. *Proceedings of the National Academy of Sciences* 119(27):e2123533119.

Hu, M. D., K. G. Lawrence, M. Gall, C. T. Emrich, M. R. Bodkin, W. B. Jackson II, N. MacNell, R. K. Kwok, L. S. Engel, and D. P. Sandler. 2021. Natural hazards and mental health among US Gulf Coast residents. *Journal of Exposure Science and Environmental Epidemiology* 31(5):842–851.

Hu, T., Y. Sun, X. Zhang, S. K. Min, and Y. H. Kim. 2020. Human influence on frequency of temperature extremes. *Environmental Research Letters* 15(6):064014.

IBHS (Insurance Institute for Business & Home Safety). 2018. *Rating the states: An assessment of residential building code and enforcement systems for life safety and property protection in hurricane-prone regions, Atlantic and Gulf Coast states, March 2018.* https://ibhs.org/wp-content/uploads/member_docs/Rating-the-States-2018_IBHS.pdf (accessed February 13, 2024).

IBHS. 2021. *Rating the states: An assessment of residential building code and enforcement systems for life safety and property protection in hurricane-prone regions, Atlantic and Gulf Coast states, June 2021.* https://ibhs.org/public-policy/rating-the-states/ (accessed January 14, 2024).

IBHS. 2024. *Rating the states: An assessment of residential building code and enforcement systems for life safety and property protection in hurricane-prone regions, Atlantic and Gulf Coast states.* https://ibhs.org/public-policy/rating-the-states/ (accessed May 21, 2024).

ICC. 2024. *Who we are: About the International Code Council.* https://www.iccsafe.org/about/who-we-are/ (accessed July 10, 2024).

Ingham, V., M. R. Islam, J. Hicks, A. Lukasiewicz, and C. Kim. 2022. Definition and explanation of community disaster fatigue. In *Complex disasters: Compounding, cascading, and protracted,* edited by A. Lukasiewicz and T. O'Donnell. Disaster risk, resilience, reconstruction and recovery (DRRR) book series. London: Palgrave Macmillan. Pp. 341–361.

Ingham, V., L. Wuersch, M. R. Islam, and J. Hicks. 2023. Indicators of community disaster fatigue: A case study in the New South Wales Blue Mountains. *International Journal of Disaster Risk Reduction* 95:103831.

International Code Council. n.d. *International codes and standards.* https://global.iccsafe.org/international-codes-and-standards/ (accessed July 11, 2024).

International Code Council. 2018. *Overview of the international residential code.* https://www.iccsafe.org/products-and-services/i-codes/2018-i-codes/irc/ (accessed February 14, 2024).

IPCC (Intergovernmental Panel on Climate Change). 2021. Summary for policymakers. In *Climate change 2021: The physical science basis.* Contribution of Working Group I to the *Sixth assessment report of the Intergovernmental Panel on Climate Change,* edited by Masson-Delmotte, V., P. Zhai, A. Pirani, S. L. Connors, C. Péan, S. Berger, N. Caud, Y. Chen, L. Goldfarb, M. I. Gomis, M. Huang, K. Leitzell, E. Lonnoy, J. B. R. Matthews, T. K. Maycock, T. Waterfield, O. Yelekçi, R. Yu, and B. Zhou. Cambridge: Cambridge University Press.

IPCC. 2022. Annex II: Glossary, edited by Möller, V, J. B. R. Matthews, R. van Diemen, C. Méndez, S. Semenov, J. S. Fuglestvedt, and A. Reisinger. In *Climate change 2022: Impacts, adaptation, and vulnerability.* Contribution of Working Group II to the *Sixth assessment report of the Intergovernmental Panel on Climate Change,* edited by H.-O. Pörtner, D. C. Roberts, M. Tignor, E. S. Poloczanska, K. Mintenbeck, A. Alegría, M. Craig, S. Langsdorf, S. Löschke, V. Möller, A. Okem, and B. Rama. Cambridge: Cambridge University Press. Pp. 2897–2930.

Islam, T., A. Marshall, L. Robinson, and E. Johnson. 2022. *Socio-economic vulnerability of African Americans in the Gulf Coast counties.* The Coastal Society 22nd International Conference, New Orleans, Louisiana. https://aquadocs.org/mapping/3917/1/Islam_papers.pdf (accessed February 14, 2024).

Javeline, D., T. Kijewski-Correa, and A. Chesler. 2021. Do Perverse Incentives Encourage Coastal Vulnerability? *Natural Hazards Review* 23(1):https://doi.org/10.1061/(ASCE)NH.1527-6996.0000533.

Jay, O., A. Capon, P. Berry, C. Broderick, R. de Dear, G. Havenith, Y. Honda, R. S. Kovats, W. Ma, A. Malik, and N. B. Morris. 2021. Reducing the health effects of hot weather and heat extremes: From personal cooling strategies to green cities. *The Lancet* 398(10301):709–724.

John Hopkins University. 2024. *Coronavirus Resource Center.* https://coronavirus.jhu.edu/map.html (accessed July 11, 2024).

Joint Center for Housing Studies of Harvard University. 2021. *The state of the nation's housing.* https://www.jchs.harvard.edu/state-nations-housing-2021 (Accessed June 20, 2024).

Joseph, J., S. M. Irshad, and A. M. Alex. 2021. Disaster recovery and structural inequalities: A case study of community assertion for justice. *International Journal of Disaster Risk Reduction* 66:102555.

Kahan, J. P., M. Wu, S. Hajiamiri, and D. Knopman. 2006. *From flood control to integrated water resource management: Lessons for the Gulf Coast from flooding in other places in the last sixty years.* Santa Monica, CA: RAND Gulf States Policy Institute.

Karaye, I., K. W. Stone, G. A. Casillas, G. Newman, and J. A. Horney. 2019. A spatial analysis of possible environmental exposures in recreational areas impacted by Hurricane Harvey flooding, Harris County, Texas. *Environmental Management* 64(4):381–390.

Karaye, I. M., A. D. Ross, and J. A. Horney. 2020. Self-rated mental and physical health of US Gulf Coast residents. *Journal of Community Health* 45(3):598–605.

Karaye, I. M., A. D. Ross, M. Perez-Patron, C. Thompson, N. Taylor, and J. A. Horney. 2019. Factors associated with self-reported mental health of residents exposed to Hurricane Harvey. *Progress in Disaster Science* 2:100016.

Keene, D. E., and K. M. Blankenship. 2023. The affordable rental housing crisis and population health equity: A multidimensional and multilevel framework. *Journal of Urban Health* 100(6):1212–1223.

Kent, A. H., and L. Ricketts. 2021. Wealth gaps between White, Black and Hispanic families in 2019. *On the Economy Blog*, Federal Reserve Bank of St. Louis, January 5. https://www.stlouisfed.org/on-the-economy/2021/january/wealth-gaps-white-black-hispanic-families-2019 (accessed March 7, 2024).

Kiefer, M., J. Rodríguez-Guzmán, J. Watson, B. van Wendel de Joode, D. Mergler, and A. S. da Silva. 2016. Worker health and safety and climate change in the Americas: Issues and research needs. *Revista Panamericana de Salud Pública* 40(3):192–197.

Kimball, S. 2023. FEMA is down to its last $3.4 billion as Maui wildfires, Hurricane Idalia slam U.S. *CNBC*, Aug 31. https://www.cnbc.com/2023/08/31/fema-is-down-to-its-last-3point4-billion-as-maui-wildfires-storms-idalia-slam-us.html (accessed October 10, 2023).

Kirmayer, L. J., H. Kienzler, A. Hamid Afana, and D. Pedersen. 2010. Trauma and disasters in social and cultural context. In *Principles of social psychiatry* (2nd ed.), edited by C. Morgan and D. Bhugra. Hoboken, NJ: Wiley Blackwell. Pp. 155–177.

Knox, C. C., D. Goodman, R. M. Entress, and J. Tyler. 2022. Compounding disasters and ethical leadership: Case studies from Louisiana and Texas. *Public Integrity* 1–19.

Knox, J. A., J. D. Frye, J. D. Durkee, and C. M. Fuhrmann. 2011. Non-convective high winds associated with extratropical cyclones. *Geography Compass* 5(2):63–89.

Knutson, T., J. McBride, J. Chan, K. Emanuel, G. Holland, C. Landsea, I. Held, J. P. Kossin, A. K. Srivastava, and M. Sugi. 2010. Tropical cyclones and climate change. *Nature Geoscience* 3:157–163.

Knutson, T. R., M. V. Chung, G. Vecchi, J. Sun, T. -L. Hsieh, and A. J. P. Smith. 2021. Climate change is probably increasing the intensity of tropical cyclones. *Science Brief News*, March 26. https://news.sciencebrief.org/cyclones-mar2021/ (accessed September 4, 2023).

Kossin, J. P. 2018. A global slowdown of tropical-cyclone translation speed. *Nature* 558(7708):104–107.

Kousky, C. 2011. Managing natural catastrophe risk: State insurance programs in the United States. *Review of Environmental Economics and Policy* 5(1):153–171.

Kousky, C., H. Wiley, and L. Shabman. 2021. Can parametric microinsurance improve the financial resilience of low-income households in the United States? *Economics of Disasters and Climate Change* (5):301–327.

Krausmann, E., A. M. Cruz, and E. Salzano. 2017. Reducing Natech risk: Organizational measures. In *Natech risk assessment and management: Reducing the risk of natural-hazard impact on hazardous installations.* Elsevier. Pp. 227–235.

Kruczkiewicz, A., J. Klopp, J. Fisher, S. Mason, S. McClain, N. M. Sheekh, R. Moss, R. M. Parks, and C. Braneon. 2021. Opinion: Compound risks and complex emergencies require new approaches to preparedness. *Proceedings of the National Academy of Sciences* 118(19):1–5.

Kruger, J., C. F. Hinton, L. B. Sinclair, and B. Silverman. 2018. Enhancing individual and community disaster preparedness: Individuals with disabilities and others with access and functional needs. *Disability and Health Journal* 11(2):170–173.

Kuhlicke, C., A. Scolobig, S. Tapsell, A. Steinführer, and B. De Marchi. 2011. Contextualizing social vulnerability: Findings from case studies across Europe. *Natural Hazards*, 58(2):789–810.

Kuran, C. H. A., C. Morsut, B. I. Kruke, M. Krüger, L. Segnestam, K. Orru, T. O. Nævestad, M. Airola, J. Keränen, F. Gabel, S. Hansson, and S. Torpan. 2020. Vulnerability and vulnerable groups from an intersectionality perspective. *International Journal of Disaster Risk Reduction* 50:101826.

Laird, H. R., C. E. Landry, J. S. Shonkwiler, and D. R. Petrolia. 2017. *Riders on the storm: Hurricane risk and coastal insurance and mitigation decisions.* https://ssrn.com/abstract=2957192 or http://dx.doi.org/10.2139/ssrn.2957192 (accessed May 20, 2024).

Lakhani, N. 2023. "We needed to get off the grid": New Orleans' community-driven response to blackouts. *The Guardian*, November 16. https://www.theguardian.com/environment/2023/nov/16/we-needed-to-get-off-the-grid-new-orleans-responds-to-its-crises-with-community-lighthouses (accessed January 18, 2024).

Lane, R. R., G. P. Kemp, and J. W. Day. 2018. A brief history of delta formation and deterioration. In *Mississippi Delta restoration: Pathways to a sustainable future*, edited by J. W. Day and J. Erdman. New York: Springer International Publishing. Pp. 11–27.

Lee, C. C., M. Maron, and A. Mostafavi. 2022. Community-scale big data reveals disparate impacts of the Texas winter storm of 2021 and its managed power outage. *Humanities and Social Sciences Communications* 9(1):1–12.

Lee, J. Y., and S. Van Zandt. 2018. Housing tenure and social vulnerability to disasters: A review of the evidence. *Journal of Planning Literature* 34(2):156–170.

Lee, R., C. J. White, M. S. G. Adnan, J. Douglas, M. D. Mahecha, F. E. O'Loughlin, E. Patelli, A. M. Ramos, M. J. Roberts, O. Martius, and E. Tubaldi. 2024. Reclassifying historical disasters: From single to multi-hazards. *Science of the Total Environment* 912:169120.

Leppold, C., L. Gibbs, K. Block, L. Reifels, and P. Quinn. 2022. Public health implications of multiple disaster exposures. *The Lancet Public Health* 7(3):e274–e286.

Lewis, P. G. T., W. A. Chiu, E. Nasser, J. Proville, A. Barone, C. Danforth, B. Kim, J. Prozzi, and E. Craft. 2023. Characterizing vulnerabilities to climate change across the United States. *Environment International* 172:107772.

Li, D., G. Newman, T. Zhang, R. Zhu, and J. Horney. 2021. Coping with post-hurricane mental distress: The role of neighborhood green space. *Social Science and Medicine* 281:114084.

Li, Y., B. R. Spoer, T. Lampe, P. Y. Hsieh, I. S. Nelson, L. E. Thorpe, and M. N. Gourevitch. 2022. Racial/ethnic and income disparities in neighborhood-level broadband access in 766 US cities, 2017–2019. *Public Health* 217:205–211.

Lichtveld, M. 2018. Disasters through the lens of disparities: Elevate community resilience as an essential public health service. *American Journal of Public Health* 108(1):28–30.

Lieberman-Cribbin, W., B. Liu, P. Sheffield, R. Schwartz, and E. Taioli. 2021. Socioeconomic disparities in incidents at toxic sites during Hurricane Harvey. *Journal of Exposure Science & Environmental Epidemiology* 31(3):454–460.

Lindsey, R., and L. Dahlman. 2023. *Climate change: Global temperature.* U.S. Department of Commerce, January 18. https://www.climate.gov/news-features/understanding-climate/climate-change-global-temperature (accessed January 14, 2024).

Lipka, M., and E. Shearer. 2023. *Audiences are declining for traditional news media in the U.S.— with some exceptions.* Pew Research Center. https://www.pewresearch.org/short-reads/2023/11/28/audiences-are-declining-for-traditional-news-media-in-the-us-with-some-exceptions/ (accessed June 6, 2024).

Lorenzini, J. A., G. Wong-Parodi, and D. R. Garfin. 2024. Associations between mindfulness and mental health after collective trauma: Results from a longitudinal, representative, probability-based survey. *Anxiety, Stress, & Coping* 37(3):361–378.

Loughran, K., and J. R. Elliott. 2019. Residential buyouts as environmental mobility: Examining where homeowners move to illuminate to illuminate social inequalities in climate adaptation. *Population Environment* 41:52–70.

Louisiana Department of Health. 2023. *LDH updates heat-related deaths during heat emergency.* https://ldh.la.gov/news/7134 (accessed February 5, 2024).

Louisiana Governor's Office of Homeland Security and Emergency Preparedness. 2012. *Hurricane Isaac: After action report and improvement plan.* https://www.yumpu.com/en/document/view/24629960/hurricane-isaac-after-action-report-lessons-learned-information- (accessed January 24, 2024).

Lowe, S., J. McGrath, M. Young, R. Kwok, L. Engel, S. Galea, and D. Sandler. 2020. Cumulative disaster exposure and mental and physical health symptoms among a large sample of residents of the U.S. Gulf Coast residents. *Journal of Traumatic Stress* 32(2):196–205.

Lowe, S. R., D. S. Fink, F. H. Norris, and S. Galea. 2015. Frequencies and predictors of barriers to mental health service use: A longitudinal study of Hurricane Ike survivors. *Social Psychiatry and Psychiatric Epidemiology* 50(1):99–108.

Lowe, S. R., S. Joshi, R. H. Pietrzak, S. Galea, and M. Cerdá. 2015. Mental health and general wellness in the aftermath of Hurricane Ike. *Social Science and Medicine* 124:162–170.

Lowe, S. R., F. H. Norris, and S. Galea. 2016. Mental health service utilization among natural disaster survivors with perceived need for services. *Psychiatric Services* 67(3):354–357.

Lowe, S. R., M. Tracy, M. Cerdá, F. H. Norris, and S. Galea S. 2013. Immediate and longer-term stressors and the mental health of Hurricane Ike survivors. *Journal of Traumatic Stress* 26(6):753–761.

Lukasiewicz, A., and T. O'Donnell. 2022. The evolution of complex disasters. In *Complex disasters: Disaster risk, resilience, reconstruction and recovery*, edited by A. Lukasiewicz and T. O'Donnell. Singapore: Palgrave Macmillan.

Lurie, N., T. Manolio, A. P. Patterson, F. Collins, and T. Frieden. 2013. Research as a part of public health emergency response. *New England Journal of Medicine* 368(13):1251–1255.

Machlis, G. E., and M. K. McNutt. 2010. Scenario-building for the *Deepwater Horizon* oil spill. *Science* 329(5995):1018–1019.

Machlis, G. E., M. O. Román, and S. T. Pickett. 2022. A framework for research on recurrent acute disasters. *Science Advances* 8(10):eabk2458.

Manisalidis, I., E. Stavropoulou, A. Stavropoulos, and E. Bezirtzoglou. 2020. Environmental and health impacts of air pollution: A review. *Frontiers in Public Health* 8(14):103389.

Marino, E. K., and A. J. Faas. 2020. Is vulnerability an outdated concept? After subjects and spaces. *Annals of Anthropological Practice* 44:33–46.

Marsooli, R., N. Lin, K. Emanuel, and K. Feng. 2019. Climate change exacerbates hurricane flood hazards along US Atlantic and Gulf Coasts in spatially varying patterns. *Nature Communications* 10:1–9.

Martin, S. A. 2015. A framework to understand the relationship between social factors that reduce resilience in cities: Application to the city of Boston. *International Journal of Disaster Risk Reduction* 12:53–80.

Martinkova, M., and J. Kysely. 2020. Overview of observed Clausius-Clapeyron scaling of extreme precipitation in midaltitudes. *Atmosphere* 11(8):786.

Maschke, A. 2023. Hurricane-battered Lake Charles dreams big on affordable housing. But is it too big? *The Advocate*, May 15. https://www.theadvocate.com/lake_charles/hurricane-battered-lake-charles-has-big-housing-plans/article_55f4497e-f10d-11ed-a1a7-cf266aa37f47.html (accessed October 20, 2023).

Masozera, M., M. Bailey, and C. Kerchner. 2007. Distribution of impacts of natural disasters across income groups: A case study of New Orleans. *Ecological Economics* 63(2–3):299–306.

Masterson, J. H., W. G. Peacock, S. S. Van Zandt, H. Grover, L. F. Schwarz, and J. T. Cooper, Jr. 2014. *Planning for community resilience: A handbook for reducing vulnerability to disasters*. Washington, DC: Island Press.

McEwen, B. S., and E. Stellar. 1993. Stress and the individual. *Archives of Internal Medicine* 153(18):2093–2101.

McEwen, L., J. Garde-Hansen, A. Holmes, O. Jones, and F. Krause. 2017. Sustainable flood memories, lay knowledges and the development of community resilience to future flood risk. *Transactions of the British Institute of Geographers* 42(1):14–28.

McKibben, J. B., C. S. Fullerton, R. J. Ursano, D. B. Reissman, K. Kowalski-Trakofler, J. M. Shultz, and L. Wang. 2010. Sleep and arousal as risk factors for adverse health and work performance in public health workers involved in the 2004 Florida hurricane season. *Disaster Medicine and Public Health Preparedness* 4(Suppl 1):S55–S62.

McKinsey and Company. 2020. *Summary: Housing study for Calcasieu Parish*. Community Foundation of Southwest Louisiana. https://www.foundationswla.org/mckinsey-housing-study (accessed January 23, 2024).

Mehta, A., M. Brennan, and J. Steil. 2020. Affordable housing, disasters, and social equity: LIHTC as a tool for preparedness and recovery. *Journal of the American Planning Association* 86(1):75–88.

Mendez, M., and J. A. Horney. 2023. Impact of the COVID-19 pandemic response on the Emergency Medical Services workforce, Texas, USA. *Journal of Emergency Management* 21(3):215–222.

Mendonça, D., T. Kijewski-Correa, A.-M. Esnard, J. Ramirez, and J. F. Olson. 2023. A framework for transitions in the built environment: Insights from compound hazards in the COVID-19 era. *Journal of Infrastructure Systems* 30(1).

Merriam-Webster Dictionary. n.d. *Epoch.* https://www.merriam-webster.com/dictionary/epoch#:~:text=%3A%20an%20extended%20period%20of%20time,a%20memorable%20series%20of%20events (accessed July 10, 2024).

Mfitumukiza, D., A. S. Roy, B. Simane, A. Hammill, M. F. Rahman, and S. Huq. 2020. *Scaling local and community-based adaptation.* Global Commission on Adaptation Background Paper. https://gca.org/reports/scaling-local-community-based-adaptation/ (accessed June 13, 2024).

Miller, H. 2020. Reopening America: A state-by-state breakdown of the status of coronavirus restrictions. *CNBC*, April 30. https://www.cnbc.com/2020/04/30/coronavirus-states-lifting-stay-at-home-orders-reopening-businesses.html (accessed August 30, 2023).

Misick, B. J. 2021. Major disaster declaration unlikely for tornado damage in southwest Louisiana, governor says. *New Orleans Public Radio*, October 28. https://www.wwno.org/news/2021-10-28/gov-edwards-major-disaster-declaration-unlikely-for-tornado-that-swept-through-sw-louisiana (accessed May 15, 2024).

Mitchell, D., and C. C. Knox. 2021. Adding insult to injury: The fiscal impact of a failed FEMA disaster reimbursement system upon Florida municipalities. *Journal of Emergency Management* 19(2):117–129.

Mitra, M., L. Long-Bellil, I. Moura, A. Miles, and H. S. Kaye. 2022. Advancing health equity and reducing health disparities for people with disabilities in the United States. *Health Affairs* 41(10):1379–1386.

Mosbergen, D. 2017. Half of Hurricane Harvey victims say FEMA application was denied or is still pending. *Huffpost*, December 6. https://www.huffpost.com/entry/harvey-fema-application-denied-survey_n_5a27a8cae4b02d3bfc368aab (accessed July 19, 2023).

Moser Design Group. n.d. *Katrina Cottages.* https://www.moserdesigngroup.com/katrina-cottages (accessed March 11, 2024).

Motairek, I., S. Rajagopalan, and S. Al-Kindi. 2023. The "heart" of environmental justice. *American Journal of Cardiology* 189:148–149.

Muller, R. A., and G. W. Stone. 2001. A climatology of tropical storm and hurricane strikes to enhance vulnerability prediction for the southeast US coast. *Journal of Coastal Research* 17:949–956.

Multi-Hazard Mitigation Council. 2019. *Natural hazard mitigation saves: 2019 report.* Washington, DC: National Institute of Building Sciences.

Muñoz, C. E., and E. Tate. 2016. Unequal recovery? Federal resource distribution after a Midwest flood disasters. *International Journal of Environmental Research and Public Health* 13(5):507.

Murley, V. A., J. D. Durkee, J. M. Gilliland, and A. W. Black. 2021. A climatology of convective and non-convective high-wind events across the eastern United States during 1973–2015. *International Journal of Climatology* 41:E368–E379.

Muyskens, J., A. Ba Tran, A. Phillips, S. Ducroquet, and N. Ahmed. 2022. More dangerous heat waves are on the way: See the impact by zip code. *The Washington Post*, August 15. https://www.washingtonpost.com/climate-environment/interactive/2022/extreme-heat-risk-map-us (accessed January 17, 2024).

Nakashima, D., I. Krupnik, and J. T. Rubis, eds. 2018. *Indigenous knowledge for climate change assessment and adaptation.* Cambridge: Cambridge University Press.

Naqvi, H. R., Y. O. Khaniabadi, D. F. Naqvi, T. C. Machathoibi, A. Shakeel, M. A. Siddiqui, P. Sicard, and A. R. Naqvi. 2023. Impact of Hurricane Ida on COVID-19 surge and its relationship with vaccination status in the United States. *Physics and Chemistry of the Earth, Parts A/B/C* 132:103469.

Nardone, A., K. E. Rudolph, R. Morello-Frosch, and J. A. Casey. 2021. Redlines and greenspace: The relationship between historical redlining and 2010 greenspace across the United States. *Environmental Health Perspectives* 129(1):017006.

NASEM (National Academies of Sciences, Engineering, and Medicine). 2015. *Healthy, resilient, and sustainable communities after disasters: Strategies, opportunities, and planning for recovery.* Washington, DC: The National Academies Press.

NASEM. 2016. *Attribution of extreme weather events in the context of climate change.* Washington, DC: The National Academies Press.

NASEM. 2017. *Communities in action: Pathways to health equity.* Washington, DC: The National Academies Press.

NASEM. 2018. *Permanent supportive housing: Evaluating the evidence for improving health outcomes among people experiencing chronic homelessness.* Washington, DC: The National Academies Press.

NASEM. 2019a. *Building and measuring community resilience: Actions for communities and the Gulf Research Program.* Washington, DC: The National Academies Press.

NASEM. 2019b. *Integrating social care into the delivery of health care: Moving upstream to improve the nation's health.* Washington, DC: The National Academies Press.

NASEM. 2020. *A framework for assessing mortality and morbidity after large-scale disasters.* Washington, DC: The National Academies Press. https://doi.org/10.17226/25863.

NASEM. 2021. *Enhancing community resilience through social capital and connectedness: Stronger together!* Washington, DC: The National Academies Press.

NASEM. 2022a. *Resilience for compounding and cascading events.* Washington, DC: The National Academies Press.

NASEM. 2022b. *Engaging socially vulnerable communities and communicating about climate change–related risks and hazards.* Washington, DC: The National Academies Press.

NASEM. 2022c. *Structural racism and rigorous models of social inequity: Proceedings of a workshop.* Washington, DC: The National Academies Press.

NASEM. 2022d. *Equitable and resilient infrastructure investments.* Washington, DC: The National Academies Press.

NASEM. 2023a. *Advancing health and resilience in the Gulf of Mexico Region: Roadmap for progress.* Washington, DC: The National Academies Press.

NASEM. 2023b. *Advancing Antiracism, Diversity, and Equity Inclusion in STEMM Organizations: Beyond Broadening Participation.* Washington, DC: The National Academies Press. https://doi.org/10.17226/26803.

NASEM. 2023c. *Strengthening equitable community resilience: Criteria and guiding principles for the Gulf Research Program's Enhancing Community Resilience (EnCoRe) initiative.* Washington, DC: The National Academies Press. https://doi.org/10.17226/26880.

NASEM. 2024. *Community-driven relocation: Recommendations for the U.S. Gulf Coast region and beyond.* Washington, DC: The National Academies Press.

National Low Income Housing Coalition. 2023. *Out of reach: The high cost of housing.* https://nlihc.org/oor (accessed June 6, 2024).

National Weather Service. n.d. *The Enhanced Fujita Scale (EF Scale).* https://www.weather.gov/oun/efscale (accessed January 24, 2024).

NCEI (National Centers for Environmental Information). n.d.-a. *Billion-dollar weather and climate disasters.* https://www.ncei.noaa.gov/access/billions/ (accessed September 7, 2023).

NCEI. n.d.-b. *Summary stats.* https://www.ncei.noaa.gov/access/billions/summary-stats/GCS/1980-2023 (accessed March 6, 2024).

NCEI. n.d.-c. *United States summary.* https://www.ncei.noaa.gov/access/billions/state-summary/US (accessed September 7, 2023).

NCEI. 2023. *Monthly tornadoes report for annual 2021.* https://www.ncei.noaa.gov/access/monitoring/monthly-report/tornadoes/202113 (accessed March 8, 2024).

Nejat, A., L. Solitare, E. Pettitt, and H. Mohsenian-Rad. 2022. Equitable community resilience: The case of Winter Storm Uri in Texas. *International Journal of Disaster Risk Reduction* 77:103070.

Nelan, M. M., and R. L. Schumann III. 2018. Gathering places in the aftermath of Hurricane Harvey. *Disaster Prevention and Management* 27(5):508–522.

Neumayer, E., and T. Plümper. 2007. The gendered nature of natural disasters: The impact of catastrophic events on the gender gap in life expectancy, 1981–2002. *Annals of the Association of American Geographers* 97(3):551–566.

Nguyen, A. M., Y. Kim, and D. M. Abramson. 2023. Neighborhood socioeconomic status and women's mental health: A longitudinal study of Hurricane Katrina survivors, 2005–2015. *International Journal of Environmental Research and Public Health* 20(2):925.

Nicole, W. 2021. A different kind of storm: Natech events in Houston's fenceline communities. *Environmental Health Perspectives* 129(5):101289.

NIHHIS (National Integrated Heat Health Information System). n.d. *About Heat.gov.* https://www.heat.gov/pages/about-heat-gov (accessed June 18, 2024).

NIST. 2020. *Community Resilience Planning Guide for Buildings and Infrastructure Systems: A Playbook.* https://nvlpubs.nist.gov/nistpubs/SpecialPublications/NIST.SP.1190GB-16.pdf (accessed July 10, 2024).

NOAA (National Oceanic and Atmospheric Administration). n.d.-a. *NOAA historical hurricane tracks: Exploring more than 150 years of historical hurricane landfalls.* https://oceanservice.noaa.gov/news/historical-hurricanes (accessed October 30, 2023).

NOAA. n.d.-b. *Gulf of Mexico atlas: Oil and gas structures.* https://www.ncei.noaa.gov/maps/gulf-data-atlas/atlas.htm?plate=Offshore%20Structures (accessed February 6, 2024).

NOAA. n.d.-c. *Gulf of Mexico atlas: Oil and gas pipelines.* https://www.ncei.noaa.gov/maps/gulf-data-atlas/atlas.htm (accessed February 6, 2024).

NOAA. 2019. The science behind the polar vortex: You might want to put on a sweater. *News & Features*, January 29. https://www.noaa.gov/multimedia/infographic/science-behind-polar-vortex-you-might-want-to-put-on-sweater (accessed January 24, 2024).

NOAA. 2020. *Annual 2020 tornadoes report.* https://www.ncei.noaa.gov/access/monitoring/monthly-report/tornadoes/202013 (accessed September 15, 2023).

NOAA. 2021a. Assessing the U.S. climate in 2021: Contiguous U.S. ranked fourth warmest during 2021; 20 billion-dollar disasters identified. National Centers for Environmental Information, *News*, January 10. https://www.ncei.noaa.gov/news/national-climate-202112 (accessed August 2, 2023).

NOAA. 2021b. Double-dip La Nina emerges: Climate pattern may influence remainder of hurricane season, winter ahead. *News & Features*, October 14. https://www.noaa.gov/news/double-dip-la-nina-emerges (accessed September 7, 2023).

NOAA. 2021c. *National Hurricane Center tropical cyclone report: Hurricane Hanna.* https://www.nhc.noaa.gov/data/tcr/AL082020_Hanna.pdf (accessed September 14, 2023).

NOAA. 2021d. *National Hurricane Center tropical cyclone report: Hurricane Laura.* https://www.nhc.noaa.gov/data/tcr/AL132020_Laura.pdf (accessed September 14, 2023).

NOAA. 2021e. *National Hurricane Center tropical cyclone report: Hurricane Sally.* https://www.nhc.noaa.gov/data/tcr/AL192020_Sally.pdf (accessed September 14, 2023).

NOAA. 2021f. *National Hurricane Center tropical cyclone report: Tropical Storm Beta.* https://www.nhc.noaa.gov/data/tcr/AL222020_Beta.pdf (accessed September 14, 2023).

NOAA. 2021g. *National Hurricane Center tropical cyclone report: Hurricane Delta.* https://www.nhc.noaa.gov/data/tcr/AL262020_Delta.pdf (accessed September 14, 2023).

NOAA. 2021h. *National Hurricane Center tropical cyclone report: Hurricane Zeta.* https://www.nhc.noaa.gov/data/tcr/AL282020_Zeta.pdf (accessed September 14, 2023).

NOAA. 2021i. *National Hurricane Center tropical cyclone report: Hurricane Eta.* https://www.nhc.noaa.gov/data/tcr/AL292020_Eta.pdf (accessed September 14, 2023).

NOAA. 2021j. *National Hurricane Center tropical cyclone report: Hurricane Nicholas.* https://www.nhc.noaa.gov/data/tcr/AL142021_Nicholas.pdf (accessed January 24, 2024).

NOAA. 2021k. *National Hurricane Center tropical cyclone report: Hurricane Fred.* https://www.nhc.noaa.gov/data/tcr/AL062021_Fred.pdf (accessed September 14, 2023).

NOAA. 2021l. *Tropical Storm Mindy—2021.* https://www.weather.gov/tae/tropicalstorm_mindy2021 (accessed January 24, 2024).

NOAA. 2021m. Record-breaking Atlantic hurricane season draws to an end. *News & Features*, updated June 10. https://www.noaa.gov/media-release/record-breaking-atlantic-hurricane-season-draws-to-end (accessed February 13, 2024).

NOAA. 2021n. It's official: July was Earth's hottest month on record. *News & Features*, August 13. https://www.noaa.gov/news/its-official-july-2021-was-earths-hottest-month-on-record (accessed December 22, 2024).

NOAA. 2021o. *February 2021 national climate report.* https://www.ncei.noaa.gov/access/monitoring/monthly-report/national/202102#SnippetTab (accessed March 7, 2024).

NOAA. 2022a. *National Hurricane Center tropical cyclone report: Hurricane Ida.* https://www.nhc.noaa.gov/data/tcr/AL092021_Ida.pdf (accessed January 19, 2024).

NOAA. 2022b. 2021 U.S. billion dollar weather and climate disasters in historical context. *Beyond the Data*, January 24. https://www.climate.gov/news-features/blogs/beyond-data/2021-us-billion-dollar-weather-and-climate-disasters-historical# (accessed October 12, 2023).

NOAA. 2022c. *National sea level rise technical report.* https://oceanservice.noaa.gov/hazards/sealevelrise/sealevelrise-tech-report.html (accessed January 18, 2024).

NOAA. 2022d. *National Hurricane Center tropical cyclone report: Tropical Storm Claudette.* https://www.nhc.noaa.gov/data/tcr/AL032021_Claudette.pdf (accessed September 14, 2023).

NOAA. 2022e. *Coastal flooding and inundation information and services at climate timescales to reduce risk and improve resilience.* Washington, DC: U.S. Department of Commerce.

NOAA. 2023. The ongoing marine heat waves in U.S. waters, explained. *News & Features,* July 14. https://www.noaa.gov/news/ongoing-marine-heat-waves-in-us-waters-explained (accessed June 10, 2024).

NOAA. 2024. *National hurricane center experimental tropical cyclone forecast cone graphic.* https://www.nhc.noaa.gov/experimental/cone/ (accessed June 1, 2024).

Nogueira, L. M., L. Sahar, J. A. Efstathiou, A. Jemal, and Y. R. Yabroff. 2019. Association between declared hurricane disasters and survival of patients with lung cancer undergoing radiation treatment. *Journal of the American Medical Association* 322(3):269–271.

Nogueira, L. M., K. R. Yabroff, and A. Bernstein. 2020. Climate change and cancer. *CA: A Cancer Journal for Clinicians* 70(4):239–244.

Nomura, S., A. J. Q. Parsons, M. Hirabayashi, R. Kinoshita, Y. Liao, and S. Hodgson. 2016. Social determinants of mid- to long-term disaster impacts on health: A systematic review. *International Journal of Disaster Risk Reduction* 16:53–67.

Norwegian Refugee Council. 2019. *Global report on internal displacement.* https://www.internal-displacement.org/sites/default/files/publications/documents/2019-IDMC-GRID.pdf (accessed January 30, 2024).

NPR (National Public Radio). 2021. Why FEMA is unavailable to many who need it the most. *All Things Considered,* June 29. https://www.npr.org/2021/06/29/1004347023/why-fema-aid-is-unavailable-to-many-who-need-it-the-most (accessed January 18, 2024).

NPR. 2023. New Orleans neighbors create spaces that can operate off the grid after hurricanes. *All Things Considered,* June 2. https://www.npr.org/2023/06/02/1179850112/new-orleans-neighbors-create-spaces-that-can-operate-off-the-grid-after-hurrican (accessed January 29, 2023).

NRC (National Research Council). 2012. *Disaster resilience: A national imperative.* Washington, DC: The National Academies Press.

NRC. 2013. *Climate and social stress: Implications for security analysis.* Washington, DC: The National Academies Press.

ODPHP (Office of Disease Prevention and Health Promotion). n.d. Social determinants of health. *Healthy People 2030.* U.S. Department of Health and Human Services. https://health.gov/healthypeople/objectives-and-data/social-determinants-health (accessed September 20, 2023).

OECD (Organisation for Economic Co-operation and Development). 2024. *Risks from natural hazards at hazardous installations (Natech). Chemical accident prevention, preparedness and response.* https://www.oecd.org/chemicalsafety/chemical-accidents/risks-from-natural-hazards-at-hazardous-installations.htm (accessed June 11, 2024).

Office of Community Development, State of Louisiana. 2024. *Action Plan Substantial Amendment No. 5.* https://cdn2.assets-servd.host/utopian-bustard/production/2020-21-Storms-APA-5-6.7.2024.pdf?dm=1717773569 (accessed June 17, 2024).

O'Keefe, P., K. Westgate, and B. Wisner. 1976. Taking the naturalness out of natural disasters. *Nature* 250:566–567.

Oliver-Smith, A. 2022. The social construction of disaster: Economic anthropological perspectives on the COVID-19 pandemic. *Economic Anthropology* 9:167–171. https://doi.org/10.1002/sea2.12236.

Oppenheimer, M., M. Campos, R. Warren, J. Birkmann, G. Luber, B. O'Neill, and K. Takahashi. 2014. Emergent risks and key vulnerabilities. In *Climate change 2014: Impacts, adaptation, and vulnerability. Part A: Global and sectoral aspects.* Contribution of Working Group II to the *Fifth assessment report of the Intergovernmental Panel on Climate Change,* edited by C. B. Field, V. R. Barros, D. J. Dokken, K. J. Mach, M. D. Mastrandrea, T. E. Bilir, M. Chatterjee, K. L. Ebi, Y. O. Estrada, R. C. Genova, B. Girma, E. S. Kissel, A. N. Levy, S. MacCracken, P. R. Mastrandrea, and L. L. White. Cambridge: Cambridge University Press. Pp. 1039–1099.

Orton, G., L. M. Fischer, and C. Lawson. 2022. Examining the impact of disaster experience with Winter Storm Uri and climate change risk perceptions on support for mitigation policy. *Journal of Applied Communications* 106(4):104148.

Osofsky, H. J., J. D. Osofsky, and T. C. Hansel. 2011. *Deepwater Horizon* oil spill: Mental health effects on residents in heavily affected areas. *Disaster Medicine and Public Health Preparedness* 5(4):280–286.

Osofsky, H., J. Osofsky, L. Saltzman, E. Lightfoot, J. De King, and T. Hansel. 2022. Mechanisms of recovery: Community perceptions of change and growth following multiple disasters. *Frontiers in Psychology* 13:1664-1078.

Ouimette, P., and J. P. Read. 2014. *Trauma and substance abuse: Causes, consequences, and treatment of comorbid disorders.* Washington, DC: American Psychological Association Press. https://psycnet.apa.org/doi/10.1037/10460-000 (accessed January 17, 2024).

Patrick, C., V. Upadhyay, A. Lucas, and K. M. Mallela. 2022. Biophysical fitness landscape of the SARS-CoV-2 Delta variant receptor binding domain. *Journal of Molecular Biology* 434(13):167622.

Patricola, C. M., and M. F. Wehner. 2018. Anthropogenic influences on major tropical cyclone events. *Nature* 563:339–346.

Peacock, W., S. Van Zandt, Y. R. Zhang, and W. E. Highfield. 2014. Inequity in long-term housing recovery after disasters. *Journal of the American Planning Association* 80(4):356–371.

Pescaroli, G., and D. E. Alexander. 2016. Critical infrastructure, panarchies and the vulnerability paths of cascading disasters. *Natural Hazards* 82, 175–192.

Pescaroli, G., and D. Alexander. 2018. Understanding compound, interconnected, interacting, and cascading risks: A holistic framework. *Risk Analysis* 38(11):2245–2257.

Pfender, E. J., K. W. Stone, K. W. Kintziger, M. A. Jagger, and J. A. Horney. 2022. Anxiety and depression among public health workers during the COVID-19 pandemic. *Journal of Emergency Management* 20(9):19–26.

Phillips, A. 2018. *Preparing for the next storm: Learning from the man-made environmental disasters that followed Hurricane Harvey.* Environmental Integrity Project. https://www.environmentalintegrity.org/wp-content/uploads/2018/08/Hurricane-Harvey-Report-Final.pdf (accessed May 29, 2024).

Phipps, S. 2022. What exactly is "the great resignation"? Middle Georgia University, *News,* April 21. https://www.mga.edu/news/2022/04/what-is-the-great-resignation.php (accessed January 24, 2024).

Pielke, R. A., Jr., and C. W. Landsea. 1998. Normalized hurricane damages in the United States: 1925–1995. *Weather and Forecasting* 13:621–631.

Pieper K. J., C. N. Jones, W. J. Rhoads, M. Rome, D. M. Gholson, A. Katner, D. E. Boellstorff, and R. E. Beighley. 2021. Microbial contamination of drinking water supplied by private wells after Hurricane Harvey. *Environmental Science and Technology* 55(12):8382–8392.

Pietrzak, R. H., M. Tracy, S. Galea, D. G. Kilpatrick, K. J. Ruggiero, J. L. Hamblen, S. M. Southwick, and F. H. Norris. 2012. Resilience in the face of disaster: Prevalence and longitudinal course of mental disorders following Hurricane Ike. *PLOS ONE* 7(6):e38964.

Potutan, G., and M. Arakida. 2021. Evolving disaster response practices during COVID-19 pandemic. *International Journal of Environmental Research and Public Health* 18(6): 3137.

Priest, A. A. 2023. Under pressure: Social capital and trust in government after natural disasters. *Social Currents* 10(4):381–400.

Priest, A. A., and J. R. Elliott. 2023. The multiplicity of impact: How social marginalization compounds climate disasters. *Environmental Sociology* 9(3):269–283.

Prochaska, J. D., A. B. Nolen, H. Kelley, K. Sexton, S. H. Linder, and J. Sullivan. 2014. Social determinants of health in environmental justice communities: Examining cumulative risk in terms of environmental exposures and social determinants of health. *Human and Ecological Risk Assessment* 20(4):980–994.

Putnam, R., 1993. The prosperous community: Social capital and public life. *The American Prospect* 4:35–42.

Putnam, R. D., and K. A. Goss. 1995. Bowling alone: America's declining social capital. *Journal of Democracy* 6(1):65–78.

Raghupathi, V., and W. Raghupathi. 2020. The influence of education on health: An empirical assessment of OECD countries for the period 1995–2015. *Archives of Public Health* 78:20.

Raker, E. J., S. R. Lowe, M. C. Arcaya, S. T. Johnson, J. Rhodes, and M. C. Waters. 2019. Twelve years later: The long-term mental health consequences of Hurricane Katrina. *Social Science & Medicine* 242:112610.

Randlett, J. A. 2010. Fair access to insurance requirements: Do "fair" property insurance premiums for individual coastal property owners in Massachusetts equate with fairness to the greater market? *Ocean & Coastal Law Journal* 15(1):127–159.

Rappaport, E. N. 2000. Loss of life in the United States associated with recent Atlantic tropical cyclones. *American Meteorological Society* (September):2065–2074.

Rappaport, E. N. 2014. Fatalities in the United States from Atlantic tropical cyclone: New data and interpretation. *American Meteorological Society* (March):341–346.

Raun, L., K. Pepple, D. Hoyt, D. Richner, A. Blanco, and J. Li. 2013. Unanticipated potential cancer risk near metal recycling facilities. *Environmental Impact Assessment Review* 41:70–77.

Reed, K. A., M. F. Wehner, and C. M. Zarzycki. 2022. Attribution of 2020 hurricane season extreme rainfall to human-induced climate change. *Nature Communications* 13:1905.

Reesman, C. 2022. *Changes in heat metrics following a major hurricane and implications on heat stress.* https://repository.lsu.edu/gradschool_theses/5580/ (accessed June 17, 2024).

Rigsby, E. A. 2016. *Understanding exclusionary zoning and its impact on concentrated poverty.* The Century Foundation, June 23. https://tcf.org/content/facts/understanding-exclusionary-zoning-impact-concentrated-poverty/?agreed=1 (accessed January 12, 2024).

Ritchie, L. A., and M. A. Long. 2021. Psychosocial impacts of post-disaster compensation processes: Community-wide avoidance behaviors. *Social Science & Medicine* 270:113640.

Ritchie, L. A., D. A. Gill, and K. Hamilton. 2022. Winter Storm Uri: Resource loss and psychosocial outcomes of critical infrastructure failure in Texas. *Journal of Critical Infrastructure Policy* 3(1):83.

Rodin, J., 2014. *The resilience dividend: Being strong in a world where things go wrong.* New York: PublicAffairs.

Rodriguez, J. M., A. S. Karlamangla, T. L. Gruenewald, D. Miller-Martinez, S. S. Merkin, and T. E. Seeman. 2019. Social stratification and allostatic load: Shapes of health differences in the MIDUS study in the United States. *Journal of Biosocial Science* 51(5):627–644.

Roque, A. D., S. H. Shah, F. Tormos-Aponte, and E. Q. Torres. 2022. *Social capital, community health resilience, and compounding hazards in Corcovada, Puerto Rico.* Natural Hazards Center, University of Colorado Boulder. https://hazards.colorado.edu/public-health-disaster-research/social-capital-community-health-resilience-and-compounding-hazards-in-corcovada-puerto-rico (accessed February 4, 2024).

Rothstein, R. 2017. *The color of law.* New York: Liveright Publishing Corporation.

Roueche, D., S. Kameshwar, J. Marshall, N. Mashrur, T. Kijewski-Correa, K. Gurley, I. Afanasyeva, G. Brasic, J. Cleary, D. Golovichev, O. Lafontaine, F. Lombardo, L. Micheli, B. Phillips, D. Prevatt, I. Robertson, J. Schroeder, D. Smith, S. Strader, M. Wilson, K. Ambrose, H. Rawajfih, and L. Rodriguez. 2020. *Hybrid preliminary virtual reconnaissance report—Early access reconnaissance report.* PVRR-EARR. StEER - Hurricane Laura. DesignSafe-CI. https://doi.org/10.17603/ds2-ng93-se16.

Rowell, D., and L. B. Connelly. 2012. A history of the term "moral hazard." *Journal of Risk and Insurance* 79(4)1051–1075.

Roy, M., and P. Berke. 2022. Social equity, land use planning, and flood mitigation. In *Oxford research encyclopedia of environmental science.*

Rufat, S., E. Tate, C. T. Emrich, and F. Antolini. 2019. How valid are social vulnerability models? *Annals of the American Association of Geographers* 109(4):1131–1153.

Rumbach, A. 2023. *Research brief: How equitable are disaster recovery programs?* https://andrewrumbach.substack.com/p/research-brief-how-equitable-are (accessed September 5, 2023).

Rumbach, A., and C. Makarewicz. 2017. Affordable housing and disaster recovery: A case study of the 2013 Colorado floods. In *Coming home after disasters: Multiple dimensions of housing recovery,* edited by A. Sapat and A. Esnard. Boca Raton, FL: CRC Press, Taylor & Francis Group. Pp. 99–112.

Rumbach, A., E. Sullivan, and C. Makarewicz. 2020. Mobile home parks and disasters: Understanding risk to the third housing type in the United States. *Natural Hazards Review* 21(2):05020001.

Rung, A. L., E. Oral, E. Fontham, D. J. Harrington, E. J. Trapido, and E. S. Peters. 2019. The long-term effects of the *Deepwater Horizon* oil spill on women's depression and mental distress. *Disaster Medicine and Public Health Preparedness* 13(2):183–190.

Rural Health Information Hub. 2024. *Health professional shortage areas: Primary care, by county, April 2024.* https://www.ruralhealthinfo.org/charts/5 (accessed June 18, 2024).

Saha, R. K., R. D. Bullard, and L. T. Powers. 2024. Liquefying the Gulf Coast: A cumulative impact assessment of LNG buildout in Louisiana and Texas. *Environmental Studies Faculty Publications* 12.

Salas, R. N., J. M. Shultz, and C. G. Solomon. 2020. The climate crisis and COVID-19—A major threat to the pandemic response. *New England Journal of Medicine* 383(11):e70.

Sandifer, P. A., R.-P. Juster, T. E. Seeman, M. Y. Lichtveld, and B. H. Singer. 2022. Allostatic load in the context of disasters. *Psychoneuroendocrinology* 140:105725.

Sansom, G., C. Thompson, L. Sansom, L. Fawkes, and E. Boerlin. 2021. *Compounding Impacts of Hazard Exposures on Mental Health in Houston, TX.* 10.21203/rs.3.rs-412401/v1.

Sansom, G. T., R. Hernandez, J. N. Johnson, G. Newman, K. Atoba, J. H. Masterson, D. Davis, and L. S. Fawkes. 2023. Evaluating the impact of proximity to reported toxic release facilities and flood events on chronic health outcomes in the city of Galena Park, Texas. *Climate Risk Management* 40:100507.

Santos-Lozada, A. R., and B. M. Rivera-Reyes. 2024. Hurricane Fiona and Puerto Rico: Compounding disasters complicate postdisaster assessments. *American Journal of Epidemiology* 193(2):404–406.

Sattler, D. N., A. J. Preston, C. F. Kaiser, V. E. Olivera, J. Valdez, and S. Schlueter. 2002. Hurricane Georges: A cross-national study examining preparedness, resource loss, and psychological distress in the U.S. Virgin Islands, Puerto Rico, Dominican Republic, and the United States. *Journal of Traumatic Stress* 15:339–350.

Schaeffer, K. 2022. *Key facts about housing affordability in the U.S.* Pew Research Center, March 23. https://www.pewresearch.org/short-reads/2022/03/23/key-facts-about-housing-affordability-in-the-u-s/ (accessed November 30, 2023).

Schmidt, S., C. Kemfert, and P. Höppe. 2009. Tropical cyclone losses in the USA and the impact of climate change—A trend analysis based on data from a new approach to adjusting storm losses. *Environmental Impact Assessment Review* 29(6):359–369.

Schuessler, A., M. Brennan, A. Mehta, and J. Steil. 2022. How can governments adapt to meet affordable housing needs after disasters? *MIT Center for Real Estate Research Paper* 23:06.

Schumann, R. L. 2018. Ground truthing spatial disaster recovery metrics with participatory mapping in post-Katrina Mississippi. *Applied Geography* 99:63–76.

Seicshnaydre, S., R. Collins, C. Hill, and M. Ciardullo. 2018. *Rigging the real estate market: Segregation, inequality, and disaster risk. The New Orleans Prosperity Index: Tricentenntial Collection.* The Data Center. https://s3.amazonaws.com/gnocdc/reports/TDC-prosperity-brief-stacy-seicshnaydre-et-al-FINAL.pdf (accessed June 12, 2024).

Sengul, H., N. Santella, L. J. Steinberg, and A. M. Cruz. 2012. Analysis of hazardous material releases due to natural hazards in the United States. *Disasters* 36(4):723–743.

Seong, K., C. Losey, and D. Gu. 2021. Naturally resilient to natural hazards? Urban-rural disparities in hazard mitigation grant program assistance. *Housing Policy Debate* 32(10):190–210.

SHA (Smart Home America). 2021. *Coastal construction code supplement.* https://www.smarthomeamerica.org/assets/uploads/SHA_Coastal_Construction_Code_Supplement_2021_Version_Final.pdf (accessed January 23, 2024).

Shah, Z., J. P. Carvallo, F. C. Hsu, and J. Taneja. 2023. The inequitable distribution of power interruptions during the 2021 Texas Winter Storm Uri. *Environmental Research: Infrastructure and Sustainability* 3(2):025011.

Shapiro, L. T., D. R. Gater, Jr., Z. Espinel, J. P. Kossin, S. Galea, and J. M. Shultz. 2020. Preparing individuals with spinal cord injury for extreme storms in the era of climate change. *EClinicalMedicine* 18:100232.

Sharma, B. B., H. R. Pemberton, B. Tonui, and B. Ramos. 2022. Responding to perinatal health and services using an intersectional framework at times of natural disasters: A systematic review. *International Journal of Disaster Risk Reduction* 76:102958.

Shepherd, J. M., A. Grundstein, and T. L. Mote. 2007. Quantifying the contribution of tropical cyclones to extreme rainfall along the coastal southeastern United States. *Geophysical Research Letters* 34(23):1–5.

Shepherd, M., and B. KC. 2015. Climate change and African Americans in the USA. *Geography Compass* 9(11):579–591.

Sherrieb, K., F. H. Norris, and S. Galea. 2010. Measuring capacities for community resilience. *Social Indicators Research* 99:227–247.

Shoaf, K. I., and S. J. Rottman. 2000. The role of public health in disaster preparedness, mitigation, response, and recovery. *Prehospital and Disaster Medicine* 15(4):144–146.

Shultz, J. M. 2019. Disasters. In *Urban health*, edited by S. Galea, C. K. Ettman, and D. Vlahov. New York: Oxford University Press.

Shultz, J. M., and D. Forbes. 2014. Psychological first aid. *Disaster Health* 2(1):3–12.

Shultz, J. M., and Y. Neria. 2013. Trauma signature analysis. *Disaster Health* 1(1):4–8.

Shultz, J. M., T. Cela, L. H. Marcelin, M. Espinola, I. Heitmann, C. Sanchez, A. Jean Pierre, C. Y. Foo, K. Thompson, P. Klotzbach, Z. Espinel, and A. Rechkemmer. 2016. The trauma signature of 2016 Hurricane Matthew and the psychosocial impact on Haiti. *Disaster Health* 3(4):121–138.

Shultz, J. M., Z. Espinel, S. Galea, and D. E. Reissman. 2007. Disaster ecology: Implications for disaster psychiatry. In *Textbook of disaster psychiatry*, edited by R. J. Ursano, C. S. Fullerton, L. Weisaeth, and B. Raphael. Cambridge: Cambridge University Press. Pp. 69–96.

Shultz, J. M., C. Fugate, and S. Galea. 2020. Cascading risks of COVID-19 resurgence during an active 2020 Atlantic hurricane season. *Journal of the American Medical Association* 324(10):935–936.

Shultz, J. M., S. Galea, Z. Espinel, A. Nori-Sarma, L. T. Shapiro, K. Dimentstein, J. M. Shepherd, and L. M. Nogueira. 2024. Safeguarding medically high-risk patients from compounding disasters. *The Lancet Regional Health–Americas* 32:100714.

Shultz, J. M., S. Galea, Z. Espinel, and D. E. Reissman. 2017. Disaster ecology. In *Textbook of disaster psychiatry,* 2nd edition, edited by R. J. Ursano, C. S. Fullerton, L. Weisaeth, and B. Raphael. Cambridge: Cambridge University Press. Pp. 44–59.

Shultz, J. M., J. P. Kossin, A. Ali, V. Borowy, C. Fugate, Z. Espinel, and S. Galea. 2020. Superimposed threats to population health from tropical cyclones in the prevaccine era of COVID-19. *Lancet Planet Health* 4(11):e506–e508.

Shultz, J. M., J. P. Kossin, A. Hertelendy, F. Burkle, C. Fugate, R. Sherman, J. Bakalar, K. Berg, A. Maggioni, Z. Espinel, D. E. Sands, R. C. LaRocque, R. N. Salas, and S. Galea. 2020. Mitigating the twin threats of climate-driven Atlantic hurricanes and COVID-19 transmission. *Disaster Medicine and Public Health Preparedness* 14(4):494–503.

Shultz, J. M., Y. Neria, A. Allen, and Z. Espinel. 2013. Psychological impacts of natural disasters. In *Encyclopedia of natural hazards*, edited by P. Bobrowsky. Dordrecht, Netherlands: Springer Publishing. Pp. 779–791.

Shultz, J. M., A. Perlin, R. G. Saltzman, Z. Espinel, and S. Galea. 2020. Pandemic march: 2019 coronavirus disease's first wave circumnavigates the globe. *Disaster Medicine and Public Health Preparedness* 16:1–5.

Shultz, J. M., J. M. Shepherd, I. Kelman, A. Rechkemmer, and S. Galea. 2018. Mitigating tropical cyclone risks and health consequences: Urgencies and innovations. *Lancet Planetary Health* 2(3):e103–e104.

Shultz, J. M., E. J. Trapido, J. P. Kossin, C. Fugate, L. Nogueira, A. Apro, M. Patel, V. J. Torres, C. K. Ettman, Z. Espinel, and S. Galea. 2022. Hurricane Ida's impact on Louisiana and Mississippi during the COVID-19 Delta surge: Complex and compounding threats to population health. *The Lancet Regional Health – Americas* 12:100286.

Shultz, J. M., L. Walsh, D. R. Garfin, F. E. Wilson, and Y. Neria. 2015. The 2010 Deepwater Horizon oil spill: The trauma signature of an ecological disaster. *Journal of Behavioral Health and Services Research* 42(1):58–76.

Sillmann, J., T. Thorarinsdottir, N. Keenlyside, N. Schaller, L. V. Alexander, G. Hegerl, S. I. Seneviratne, R. Vautard, X. Zhang, and F. Zwiers. 2017. Understanding, modeling and predicting weather and climate extremes: Challenges and opportunities. *Weather and Climate Extremes* 18:65–74.

Simms, J. R. Z. 2017. "Why would I live anyplace else?": Resilience, sense of place, and possibilities of migration in coastal Louisiana. *Journal of Coastal Research* 33(2):408–420.

Simpson, N. P., K. J. Mach, A. Constable, J. Hess, R. Hogarth, M. Howden, J. Lawrence, R. J. Lempert, V. Muccione, B. Mackey, M. G. New, B. O'Neill, F. Otto, H.-O. Pörtner, A. Reisinger, D. Roberts, D. N. Schmidt, S. Seneviratne, S. Strongin, M. van Aalst, and C. H. Trisos. 2021. Assessing and responding to complex climate change risks. *One Earth* 4(4):489–501.

Singh, D., A. R. Crimmins, J. M. Pflug, P. L. Barnard, J. F. Helgeson, A. Hoell, F. H. Jacobs, M. G. Jacox, A. Jerolleman, and M. F. Wehner. 2023. Focus on compound events. In *Fifth national climate assessment*, edited by A. R. Crimmins, C. W. Avery, D. R. Easterling, K. E. Kunkel, B. C. Stewart, and T. K. Maycock. Washington, DC: U.S. Global Change Research Program.

Smiley, K. T. 2020. Social inequalities in flooding inside and outside of floodplains during Hurricane Harvey. *Environmental Research Letters* 15(9):0940b3.

Smiley, K. T., J. Howell, and J. R. Elliott. 2018. Disasters, local organizations, and poverty in the USA, 1998 to 2015. *Population and Environment* 40(2):115–135.

Smiley, K. T., I. Noy, M. F. Wehner, D. Frame, C. C. Sampson, and O. E. Wing. 2022. Social inequalities in climate change-attributed impacts of Hurricane Harvey. *Nature Communications* 13(1):3418.

Smit, B., and J. Wandel. 2006. Adaption, adaptive capacity and vulnerability. *Global Environmental Change* 16(3):282–292.

Smith, A. 2022. U.S. billion-dollar weather and climate disasters in historical context. *Beyond the Data*, January 24. https://www.climate.gov/news-features/blogs/beyond-data/2021-us-billion-dollar-weather-and-climate-disasters-historical (accessed November 21, 2023.)

Smith, M. 2022. Hurricane-hit southwest Louisiana's population drop among steepest in nation. *The Advocate*, March 25. https://www.theadvocate.com/lake_charles/hurricane-hit-southwest-louisiana-s-population-drop-among-steepest-in-nation/article_44d67698-abb4-11ec-9763-a70b7b6adfc4.html (accessed October 16, 2023).

Sobhaninia, S. 2023. Does social cohesion accelerate the recovery rate in communities impacted by environmental disasters in Puerto Rico? An analysis of a community survey. *Environmental Advances* 13:100400.

Song, F., G. J. Zhang, V. Ramanathan, and L. R. Leung. 2022. Trends in surface equivalent potential temperature: A more comprehensive metric for global warming and weather extremes. *Proceedings of the National Academy of Sciences* 119(6):e2117832119.

St. Aubin, C., and J. Liedke. 2023. *Most Americans favor restrictions on false information, violent content online.* https://www.pewresearch.org/short-reads/2023/07/20/most-americans-favor-restrictions-on-false-information-violent-content-online/ (accessed March 11, 2024).

Stanley, J. 2017. Equity in recovery. In *Urban planning for disaster recovery*, edited by A. March and M. Kornakova. Oxford, UK: Butterworth-Heinemann. Pp. 31–45.

State of Louisiana. 2022. *Proposed master action plan.* https://cdn2.assets-servd.host/utopian-bustard/production/OCD-Laura-Delta-Action-Plan-FINAL-7-28-22.pdf (accessed January 24, 2024).

State of Louisiana. 2023. The Watershed Initiative. https://watershed.la.gov (accessed January 24, 2024).

Statista. 2021. *Death rates from coronavirus (COVID-19) in the United States as of March 10, 2023, by state.* https://www.statista.com/statistics/1109011/coronavirus-covid-death-rates-us-by-state (accessed August 22, 2023).

Steinberg, T. 2000. *The unnatural history of natural disasters in America.* New York: Oxford University Press.

Stott, P. A., D. A. Stone, and M. R. Allen. 2004. Human contribution to the European heatwave of 2003. *Nature* 432:610–614.

Strader, S. M. 2023. *How a changing societal landscape is shaping Gulf Coast tropical cyclone and tornado disasters.* Paper commissioned by the Committee on Compounding Disasters in Gulf Coast Communities, 2020–2021: Impacts, Findings, and Lessons Learned, National Academies of Sciences, Engineering, and Medicine.

Strader, S. M., and W. S. Ashley. 2018. Finescale assessment of mobile home tornado vulnerability in the central and southeast United States. *Weather, Climate, and Society* 10:797–812.

Strader, S. M., W. S. Ashley, A. M. Haberlie, and K. Kaminski. 2022. Revisiting US nocturnal tornado vulnerability and its influence on tornado impacts. *Weather, Climate, and Society* 14(4):1147–1163.

Sugg, M. M., L. Wertis, S. C. Ryan, S. Green, D. Singh, and J. D. Runkle. 2023. Cascading disasters and mental health: The February 2021 winter storm and power crisis in Texas, USA. *Science of the Total Environment* 880:163231.

Sullivan, E., C. Makarewicz, A. Rumbach, and N. Albert. 2021. *Affordable but marginalized: Study provides first comprehensive look at Houston's mobile home parks.* https://kinder.rice.edu/urbanedge/affordable-marginalized-study-provides-first-comprehensive-look-houstons-mobile-home (accessed May 2, 2024).

Swope, C. B., and D. Hernández. 2019. Housing as a determinant of health equity: A conceptual model. *Social Science & Medicine* (243):112571.

TASCE (Texas Section of the American Society of Civil Engineers). 2022. *Reliability and resilience in the balance: Building resilient infrastructure for a reliable future. A vision for storms beyond Uri and Viola.* https://www.texasce.org/wp-content/uploads/2022/02/Reliability-Resilience-in-the-Balance-REPORT.pdf (accessed July 10, 2024).

Tate, E., M. A. Rahman, C. T. Emrich, and C. C. Sampson. 2021. Flood exposure and social vulnerability in the United States. *Natural Hazards* 106(1):435–457.

Taylor, K. 2019. *Race for profit: How banks and the real estate industry undermined Black homeownership.* Chapel Hill: The University of North Carolina Press.

Terrell, K. A., and W. James. 2022. Racial disparities in air pollution burden and COVID-19 deaths in Louisiana, USA, in the context of long-term changes in fine particulate pollution. *Environmental Justice* 15(5):286–297.

Terrell, K. A., and G. St. Julien. 2023. Discriminatory outcomes of industrial air permitting in Louisiana, United States. *Environmental Challenges* 10:100672.

Texas Department of State Health Services. 2021. *February 2021 winter storm-related deaths—Texas.* https://www.dshs.texas.gov/sites/default/files/news/updates/SMOC_FebWinterStorm_MortalitySurvReport_12-30-21.pdf (accessed January 24, 2024).

Texas Housers. 2018. *Low-income households disproportionately denied by FEMA is a sign of a system that is failing the most vulnerable.* https://texashousers.org/2018/11/30/low-income-households-disproportionately-denied-by-fema-is-a-sign-of-a-system-that-is-failing-the-most-vulnerable/ (accessed September 11, 2023).

Texas State Comptroller. 2021. *Winter Storm Uri 2021: The 87th Legislature takes on electricity reform.* https://comptroller.texas.gov/economy/fiscal-notes/2021/oct/winter-storm-reform.php (accessed August 15, 2023).

Theodorou, J. 2023. The truth about catastrophes. *R Street Shorts, No. 133.* https://www.rstreet.org/wp-content/uploads/2023/12/FINAL-policy-short-no-133-2.pdf (accessed January 23, 2024).

Thiede, B. C., and D. L. Brown. 2013. Hurricane Katrina: Who stayed and why? *Population Research Policy Review* 32:803–824.

Thomas, D. S. K., S. Jang, and J. Scandlyn. 2020. The CHASMS conceptual model of cascading disasters and social vulnerability: The COVID-19 case example. *International Journal of Disaster Risk Reduction* 51:101828.

Tierney, K. 2019. *Disasters: A sociological approach.* Chichester, UK: John Wiley & Sons.

Tobin, R. J., and C. Calfee. 2005. *The National Flood Insurance Program's Mandatory Purchase Requirement: Policies, Processes, and Stakeholders.* www.fema.gov/sites/default/files/2020-07/fema_nfip_eval_mandatory_purchase_requirement.pdf.

Tomko, B., C. L. Nittrouer, X. Sanchez-Vila, and A. H. Sawyer. 2023. Disparities in disruptions to public drinking water services in Texas communities during Winter Storm Uri 2021. *PLOS Water* 2(6):p.e0000137.

Törnqvist, T. E., K. L. Jankowski, Y.-X. Li, and J. L. González. 2020. Tipping points of Mississippi Delta marshes due to accelerated sea-level rise. *Science Advances* 6(21):eaaz5512.

Torres, J. M., and J. A. Casey. 2017. The centrality of social ties to climate migration and mental health. *BMC Public Health* 17(1):1–10.

Tosun, J., and M. Howlett. 2021. Managing slow onset events related to climate change: The role of public bureaucracy. *Current Opinion in Environmental Sustainability* 50:43–53.

Tozier de la Poterie, A., Y. Clatworthy, E. Easton-Calabria, E. Coughlan de Perez, S. Lux, and M. van Aalst. 2022. Managing multiple hazards: Lessons from anticipatory humanitarian action for climate disasters during COVID-19. *Climate and Development* 14(4):374–388.

Tracy, M., F. H. Norris, and S. Galea. 2011. Differences in the determinants of posttraumatic stress disorder and depression after a mass traumatic event. *Depression & Anxiety* 28(8):666–675.

Trivedi, J. 2020. *Mississippi after Katrina: Disaster recovery and reconstruction on the Gulf Coast.* Lanham, MD: Lexington Books.

Trivedi, J. 2023. *Compounding disasters in Gulf Coast communities, 2020–2021: Impacts, findings, and lessons learned in Jefferson Davis and Marion counties, Mississippi.* Paper commissioned by the Committee on Compounding Disasters in Gulf Coast Communities, 2020–2021: Impacts, Findings, and Lessons Learned, National Academies of Sciences, Engineering, and Medicine.

Tsai, S. L., C. Ochiai, M. C. Hou, and M. H. Tseng. 2021. Spatial and project planning characteristics of post-disaster settlement a case study of reconstruction after Typhoon Morakot in Taiwan. *Journal of the City Planning Institute of Japan* 56(3):936–943.

Ueland, J., and B. Warf. 2006. Racialized topographies: Altitude and race in Southern cities. *Geographical Review* 96(1):50–78.

Underhill, M. 2009. The invisible toll of Katrina: How social and economic resources are altering the recovery experience among Katrina evacuees in Colorado. In *The political economy of hazards and disasters*, edited by E. C. Jones and A. D. Murphy. Lanham, MD: Alta Mira Press. Pp. 59–82.

UNDRR (United Nations Office for Disaster Risk Reduction). n.d.-a. Vulnerability. *PreventionWeb*, Understanding Disaster Risk. https://www.preventionweb.net/understanding-disaster-risk/component-risk/vulnerability (accessed January 30, 2024).

UNDRR. n. d.-b. *Sendai framework terminology on disaster risk reduction: Hazard.* https://www.undrr.org/terminology/hazard?quickUrl=true (accessed January 18, 2024).

UNDRR. 2022. The importance of early warning systems in disaster risk reduction. *PreventionWeb*, October 13. https://www.preventionweb.net/news/importance-early-warning-systems-disaster-risk-reduction (accessed January 18, 2024).

UNHCR (United Nations High Commissioner for Refugees). 2024. Humanitarian principles. *Emergency Handbook*, last updated January 30, 2024. https://emergency.unhcr.org/protection/protection-principles/humanitarian-principles (accessed June 17, 2024).

Union of Concerned Scientists. 2014. *Overwhelming risk: Rethinking flood insurance in a world of rising seas.* https://www.ucsusa.org/sites/default/files/2019-09/Overwhelming-Risk-Full-Report.pdf (accessed March 23, 2020).

U.S. Bureau of Labor Statistics. 2023a. *Local area unemployment statistics: Unemployment rates for states.* https://www.bls.gov/lau/lastrk19.htm (accessed February 13, 2024).

U.S. Bureau of Labor Statistics. 2023b. *Local area unemployment statistics: Unemployment rates for states.* https://www.bls.gov/lau/lastrk20.htm (accessed February 13, 2024).

U.S. Bureau of Labor Statistics. 2023c. *Local area unemployment statistics: Unemployment rates for states.* https://www.bls.gov/lau/lastrk21.htm (accessed February 13, 2024).

U.S. Bureau of Labor Statistics. 2023d. *Databases, tables, and calculators by subject.* https://data.bls.gov/timeseries/LNS14000000 (accessed February 13, 2024).

U.S. Census Bureau. 2019. *Coastline America.* https://www.census.gov/content/dam/Census/library/visualizations/2019/demo/coastline-america.pdf (accessed November 6, 2023).

U.S. Census Bureau. 2020. *Vacancy Status. Decennial Census, Table H5.* https://data.census.gov/table/DECENNIALDHC2020.H5?q=H5:%20VACANCY%20STATUS&g=050XX-00US01003,01097,22019,22023,48167,48201 (accessed on May 18, 2024).

U.S. Census Bureau. 2021a. *QuickFacts—Baldwin, Mobile County, Alabama; Harris, Galveston County, Texas; Cameron, Calcasieu Parish Louisiana; Marion, Jefferson Davis County, Mississippi.* https://www.census.gov/quickfacts/fact/table/baldwincountyalabama,mobilecountyalabama,harricountytexas,galvestoncountytexas,cameronparishlouisiana,calcasieuparishlouisiana,marioncountymississippi,jeffersondaviscountymississippi (accessed February 13, 2024).

U.S. Census Bureau. 2021b. *Household income: 2021.* American Community Survey Briefs, ACSBR-011. https://www.census.gov/content/dam/Census/library/publications/2022/acs/acsbr-011.pdf (accessed January 23, 2024).

U.S. Census Bureau. 2021c. *Income and poverty in the United States: 2020.* Report no. P60-273. https://www.census.gov/library/publications/2021/demo/p60-273.html (accessed February 14, 2024).

U.S. Census Bureau. 2022a. *Glossary.* https://www.census.gov/programs-surveys/geography/about/glossary.html#:~:text=Block%20Groups%20(BGs)%20are%20statistical,data%20and%20control%20block%20numbering (accessed March 15, 2024).

U.S. Census Bureau. 2022b. Physical housing characteristics for occupied housing units. *American Community Survey*, ACS 1-Year Estimates Subject Tables, Table S2504. https://data.census.gov/table/ACSST1Y2022.S2504?q=Year Structure Built&g=040XX-00US01,12,22,28,48 (accessed on May 18, 2024).

U.S. Census Bureau. 2022c. Growth in U.S. population shows early indication of recovery amid COVID-19 pandemic. News release, December 22. https://www.census.gov/newsroom/press-releases/2022/2022-population-estimates.html (accessed January 22, 2024).

U.S. Census Bureau. 2023a. *MHS latest data.* https://www.census.gov/data/tables/time-series/econ/mhs/latest-data.html (accessed January 23, 2024).

U.S. Census Bureau. 2023b. *Poverty in the United States: 2022.* https://www.census.gov/content/dam/Census/library/publications/2023/demo/p60-280.pdf (accessed January 23, 2024).

U.S. Census Bureau. 2023c. *Persistent poverty in counties and census tracts.* https://www.census.gov/content/dam/Census/library/publications/2023/acs/acs-51%20persistent%20poverty.pdf (accessed January 23, 2024).

U.S. Census Bureau. 2024. *Household pulse survey public use file.* https://www.census.gov/programs-surveys/household-pulse-survey/datasets.html (accessed March 7, 2024).

USDA (United States Department of Agriculture). 2017. *Identifying potential heirs properties in the Southeastern United States.* https://www.srs.fs.usda.gov/pubs/gtr/gtr_srs225.pdf (accessed February 16, 2024).

USDA. 2023. *Heirs' property.* https://www.nal.usda.gov/farms-and-agricultural-production-systems/heirs-property (accessed October 6, 2023).

USDA. 2024. *Climate hubs, extreme weather.* https://www.climatehubs.usda.gov/content/extreme-weather#:~:text=Weather%2Drelated%20extreme%20events%20are,a%20longer%20period%20of%20time (accessed June 17, 2024).

USEIA (U.S. Energy Information Administration). 2023a. *Gulf of Mexico fact sheet.* https://www.eia.gov/special/gulf_of_mexico/ (accessed May 23, 2024).

USEIA. 2023b. Europe was the main destination for U.S. LNG exports in 2022. *Today in Energy*, March 22. https://www.eia.gov/todayinenergy/detail.php?id=55920# (accessed May 17, 2024).

U.S. EPA (U.S. Environmental Protection Agency). 2019. *EPA needs to improve its emergency planning to better address air quality concerns during future disasters.* Report no. 20-P-0062. https://www.epa.gov/sites/production/files/2019-12/documents/_epaoig_20191216-20-p-0062.pdf (accessed May 29, 2024).

U.S. EPA. 2021. *Climate change and social vulnerability in the United States: A focus on six impacts.* https://www.epa.gov/cira/social-vulnerability-report (accessed December 13, 2023).

U.S. GAO (U.S. Government Accountability Office). 2020. *Disaster assistance: Additional actions needed to strengthen FEMA's individuals and households program.* https://www.gao.gov/assets/gao-20-503.pdf (accessed January 24, 2024).

U.S. GAO. 2022. *Disaster recovery: Better data are needed to ensure equitable delivery of HUD block grant funds to vulnerable populations.* https://www.gao.gov/products/gao-22-105548 (accessed January 23, 2024).

USGCRP (United States Global Change Research Program). 2023. *Fifth national climate assessment*, edited by A. R. Crimmins, C. W. Avery, D. R. Easterling, K. E. Kunkel, B. C. Stewart, and T. K. Maycock. Washington, DC: U.S. Global Change Research Program. https://nca2023.globalchange.gov/ (accessed January 19, 2024).

USGS (United States Geological Survey). n.d. *Texas Gulf Coast groundwater and land subsidence: Over forty years of research in the Houston-Galveston region.* https://txpub.usgs.gov/houston_subsidence/ (accessed February 15, 2024).

U.S. Treasury Department. 2021. *State and local fiscal recovery. Allocations and payments: Allocations for counties.* https://home.treasury.gov/system/files/136/fiscalrecoveryfunds_countyfunding_2021.05.10-1a-508A.pdf (accessed January 31, 2024).

van den Hurk, B. J., C. J. White, A. M. Ramos, P. J. Ward, O. Martius, I. Olbert, K. Roscoe, H. M. Goulart, and J. Zscheischler. 2023. Consideration of compound drivers and impacts in the disaster risk reduction cycle. *iScience* 26(3):106030.

van der Wiel, K., S. B. Kapnick, G. J. van Oldenborgh, K. Whan, S. Philip, G. A. Vecchi, R. K. Singh, J. Arrighi, and H. Cullen. 2016. Rapid attribution of the August 2016 flood-inducing extreme precipitation in south Louisiana to climate change. *Hydrology and Earth System Sciences—Discussion* 21(2):897–921.

Van Zandt, S., and M. Sloan. 2017. The Texas experience with 2008's Hurricanes Dolly and Ike. In *Coming home after disaster: Multiple dimensions of housing recovery*, edited by A. Sapat and A. Esnard. Boca Raton, FL: CRC Press, Taylor & Francis Group. Pp. 83–98.

Wagner, K. 2019. Adaptation and adverse selection in markets for natural disaster insurance. *SSRN*, October 10. http://dx.doi.org/10.2139/ssrn.3467329 (accessed June 12, 2024).

Walsh, K. J., J. L. McBride, P. J. Klotzbach, S. Balachandran, S. J. Camargo, G. Holland, T. R. Knutson, J. P. Kossin, T. C. Lee, A. Sobel, and M. Sugi. 2016. Tropical cyclones and climate change. *Wiley Interdisciplinary Reviews: Climate Change* 7(1):65–89.

Wang, G., X. Zhou, K. Wang, X. Ke, Y. Zhang, R. Zhao, and Y. Bao. 2020. GOM20: A stable geodetic reference frame for subsidence, faulting, and sea-level rise studies along the coast of the Gulf of Mexico. *Remote Sensing* 12(3):350.

Wang, P. S., M. J. Gruber, R. E. Powers, M. Schoenbaum, A. H. Speier, K. B. Wells, and R. C. Kessler. 2008. Disruption of existing mental health treatments and failure to initiate new treatment after Hurricane Katrina. *American Journal of Psychiatry* 165(1):34–41.

Wang, Q., Y. Xu, N. Wei, S. Wang, and H. Hu. 2019. Forecast and service performance on rapidly intensification process of typhoons Rammasun (2014) and Hato (2017). *Tropical Cyclone Research and Review* 8(1):18–26.

Waters, M. C. 2016. Life after Hurricane Katrina: The resilience in survivors of Katrina (RISK) project. *Sociological Forum* 31:750–769.

Weden, M. M., V. Parks, A. M. Parker, L. Drakeford, and R. Ramchand. 2021. Health disparities in the U.S. Gulf Coast: The interplay of environmental disaster, material loss, and residential segregation. *Environmental Justice* 14(2):110–123.

Wells, E. M., M. Boden, I. Tseytlin, and I. Linkov. 2022. Modeling critical infrastructure resilience under compounding threats: A systematic literature review. *Progress in Disaster Science* 15:100244.

White House Council of Economic Advisers. 2022. *Economic report of the President.* https://www.whitehouse.gov/wp-content/uploads/2022/04/ERP_2022_.pdf (accessed January 24, 2024).

WHO. 2020a. Statement on the second meeting of the International Health Regulations (2005) Emergency Committee regarding the outbreak of novel coronavirus (2019-nCoV). Accessed July 10, 2024. https://www.who.int/news/item/30-01-2020-statement-on-the-second-meeting-of-the-international-health-regulations-(2005)-emergency-committee-regarding-the-outbreak-of-novel-coronavirus-(2019-ncov).

WHO. 2020b. WHO Director-General's opening remarks at the media briefing on COVID-19 - 11 March 2020. Accessed July 10, 2024. https://www.who.int/director-general/speeches/detail/who-director-general-s-opening-remarks-at-the-media-briefing-on-covid-19---11-march-2020.

Whytlaw, J. L., N. Hutton, J.-E. W. Yusuf, T. Richardson, S. Hill, T. Olanrewaju-Lasisi, P. Antwi-Nimarko, E. Landaeta, and R. Diaz. 2021. Changing vulnerability for hurricane evacuation during a pandemic: Issues and anticipated responses in the early days of the COVID-19 pandemic. *International Journal of Disaster Risk Reduction* 61:102386.

Wickerson, G. 2023. *Six hot opportunity areas to beat the heat through federal policy.* https://fas.org/publication/beat-the-heat-through-federal-policy/ (accessed June 17, 2023).

Wilbanks, T. J. 2008. Enhancing the resilience of communities to natural and other hazards: What we know and what we can do. *Natural Hazards Observer* 32:10–11.

Williams, J. 2023. Why is Mississippi experiencing more tornadoes in comparison to past years? *Clarion Ledger*, April 21. https://www.clarionledger.com/story/news/2023/04/21/climate-change-causing-more-tornadoes-in-mississippi/70133867007/ (accessed September 14, 2023).

Witze, A. 2018. Why extreme rains are gaining strength as the climate warms. *Nature*, November 20. https://www.nature.com/articles/d41586-018-07447-1 (accessed September 15, 2023).

WJTV. 2020. *Statewide burn ban in effect amid COVID-19 crisis.* https://www.wjtv.com/news/statewide-burn-ban-in-effect-amid-covid-crisis (accessed January 25, 2024).

Xi, D., and N. Lin. 2021. Sequential landfall of tropical cyclones in the United States: From historical records to climate projections. *Geophysical Research Letters* 48(21):e2021GL094826.

Xi, D., N. Lin, and A. Gori. 2023. Increasing sequential tropical cyclone hazards along the US East and Gulf Coasts. *Nature Climate Change* 13:258–265.

Xu, J. Q., S. L. Murphy, K. D. Kochanek, and E. Arias. 2022. *Mortality in the United States, 2021.* NCHS Data Brief 456. Hyattsville, MD: National Center for Health Statistics.

Yamori, K. and J. Goltz. 2021. Disasters without Borders: The Coronavirus Pandemic, Global Climate Change and the Ascendancy of Gradual Onset Disasters. International Journal of Environmental Research and Public Health 18(6): 3299. https://doi.org/10.3390/ijerph18063299.

Yin, J. 2023. Rapid decadal acceleration of sea level rise along the U.S. East and Gulf Coasts during 2010–22 and its impact on hurricane-induced storm surge. *Journal of Climate* 36(13):4511–4529.

Yu, Q., W. Cao, D. Hamer, N. Urbanek, S. Straif-Bourgeois, S. A. Cormier, T. Ferguson, and J. Richmond-Bryant. 2023. Associations of COVID-19 hospitalizations, ICU admissions, and mortality with Black and White race and their mediation by air pollution and other risk factors in the Louisiana industrial corridor, March 2020–August 2021. *International Journal of Environmental Research and Public Health* 20(5):4611.

Yzermans, C. J., G. A. Donker, J. J. Kerssens, A. J. Dirkwagner, R. J. Soetman, and M. ten Veen. 2005. Health problems of victims before and after disaster: A longitudinal study in general practice. *International Journal of Epidemiology* 34:820–826.

Zavar, E. M., and R. L. Schumann. 2019. Patterns of disaster commemoration in long-term recovery. *Geographical Review* 109(2):157–179.

Zhou, X., G. Wang, K. Wang, H. Liu, H. Lyu, and M. J. Turco. 2021. Rates of natural subsidence along the Texas coast derived from GPS and tide gauge measurements (1904–2020). *Journal of Surveying Engineering* 174(4):04021020.

Zhu, L., and S. M. Quiring. 2022. Exposure to precipitation from tropical cyclones has increased over the continental United States from 1948 to 2019. *Communications Earth & Environment* 3: Article 312.

Zscheischler, J., and F. Lehner. 2022. Attributing compound events to anthropogenic climate change. *Bulletin of the American Meteorological Society* 103(3):E936–E953.

Zscheischler, J., S. Westra, B. J. J. M. Van den Hurk, S. I. Seneviratne, P. J. Ward, A. Pitman, A. AghaKouchak, D. N. Bresch, M. Leonard, T. Wahl, and X. Zhang. 2018. Future climate risk from compound events. *Nature Climate* 8:469–477.

Appendix A

Glossary

The Committee on Compounding Disasters in Gulf Coast Communities, 2020–2021 recognizes the important contribution of terms and definitions offered previously by scholars and researchers and reflected in foundational documents, including National Academies of Sciences, Engineering, and Medicine consensus studies, and national and international reports. The committee also recognizes the continued rapid expansion of disaster research and continued refinement of definitions. In the interest of clarity, the committee presents the following as a set of consistent definitions for key terms used in this report. The committee acknowledges that these are not the only definitions available in the literature for these terms, nor are they necessarily the definitions used in all communities or by all practitioners. They are presented for the purpose of establishing a shared understanding of how terms are used in this report, and unless particularly addressed within a recommendation in this report, are not intended for any other purpose than establishing a shared understanding between the committee and the reader. The following terms are used extensively throughout this report and the associated definitions represent the consensus of the committee, unless otherwise noted in the report content.

ADAPTATION, CLIMATE: In human systems, the process of adjustment to actual or expected climate and its effects to moderate harm or exploit beneficial opportunities. In natural systems, the process of adjustment to actual climate and its effects. Human intervention may facilitate adjustment to expected climate and its effects (IPCC, 2022, Annex II).

ADAPTATION, EQUITABLE: Equitable adaptation addresses the disproportionate effects of climate change for overburdened and front-line communities. It dismantles barriers, considers underlying stresses, creates opportunities, and enables learning through iterative evaluation and sustained engagement and intentionally incorporates recognitional, procedural, contextual, and distributional principles of equity in design, planning, and execution.

ADAPTIVE CAPACITY: The ability of systems, institutions, and humans to adjust to potential damage, to take advantage of opportunities, or to respond to consequences (USGCRP, 2023).

CAPACITY: Capacity refers to all the strengths, attributes and resources available within a community, organization, or society to manage and reduce disaster risks and strengthen resilience (UNDRR, n.d.-b).

COMMUNITY: The members of a collectivity, who share a common territorial area as their base of operation for daily activities. Also, a social group whose members are bound together by the sense of belonging created out of everyday contacts covering the entire range of human activities (NASEM, 2021).

COMPOUND EVENT: An event that consists of two or more extreme events occurring simultaneously or successively, combinations of extreme events with underlying conditions that amplify the impact of the events, or combinations of events that are not themselves extremes but lead to an extreme event or impact when combined. The contributing events can be of similar or different types (USGCRP, 2023).

DISASTER: A serious disruption of the functioning of a community or a society at any scale due to hazardous events interacting with conditions of exposure, vulnerability, and capacity, leading to one or more of the following: human, material, economic, and environmental losses and impacts (IPCC, 2022).

DISASTER, CASCADING: Cascading disasters are extreme events, in which cascading effects increase in progression over time and generate unexpected secondary events of strong impact. These tend to be at least as serious as the original event, and contribute significantly to the overall duration of the disaster's effects. These subsequent and unanticipated crises can be exacerbated by the failure of physical structures, and the social functions that depend on them, including critical facilities, or by the inadequacy of disaster mitigation strategies, such as evacuation procedures, land-use planning and emergency management strategies. Cascading disasters tend to highlight unresolved vulnerabilities in human society. In cascading disasters one or more secondary events can be identified and distinguished from the original source of disaster (Pescaroli and Alexander, 2016).

DISASTER, COMPOUNDING: The result of concurrent, consecutive (de Ruiter et al., 2020), and/or overlapping disruptive events that affect the societal, governmental, and/or environmental functions of a community or region and diminish the community's capacity to recover and resume essential activities. The weakening of these interrelated functions inhibits and prolongs the disaster recovery period, making communities more likely to experience amplified negative effects of future disruptive events. Some communities are at disproportionate risk of suffering the effects of compounding disasters as a result of the interplay of persistent physical and social vulnerability factors and increased exposure to climatic and non-climatic hazards.

DISASTER RISK: Disaster risk is expressed as the likelihood of loss of life, injury or destruction and damage from a disaster in a given period of time. Disaster risk is widely recognized as the consequence of the interaction between a hazard and the characteristics that make people and places vulnerable and exposed (UNDRR, n.d.-b). Disaster risk not only depends on the severity of hazard or the number of people or assets exposed, but it is also a reflection of the susceptibility of people and economic assets to suffer loss and damage.

DISASTER RISK MANAGEMENT: Processes for designing, implementing, and evaluating strategies, policies, and measures to improve the understanding of current and future disaster risk, foster disaster risk reduction and transfer, and promote continuous improvement in disaster preparedness, prevention and protection, and response and recovery practices, with the explicit purpose of increasing human security, well-being, quality of life, and sustainable development (IPCC, 2022, Annex II).

DISRUPTIVE EVENT: The occurrence of one or more hazards that negatively affects the functioning of a community or society due to interactions with conditions of exposure, vulnerability and capacity. The occurrence of a disruptive event is time bound, though the impacts may persist indefinitely. For example, a hurricane is a disruptive event defined by specific meteorological conditions related to the organization and intensity of the storm system. While this disruptive event ends when those conditions are no longer present at a given location, perhaps within hours, the recovery from that disruptive event often extends for years or even decades.

EPOCH: An extended period of time usually characterized by a distinctive development or by a memorable series of events (Merriam-Webster Dictionary, 2024).

EQUITY: An outcome from fair conditions (policies, practices, structures, cultures, and norms) in which all individuals and groups have the opportunities and resources they need for general well-being or success in specific metrics (such as pay or advancement). Equity is aligned with justice and may require the systemic redistribution of power, access, and resources. Equity should not be confused with equality, which is the treatment of all individuals in the same manner regardless of their starting point (NASEM, 2023b).

EQUITY, PROCEDURAL: The degree to which fair treatment characterizes policies and programs (Bullard, 2005).

EQUITY, SOCIAL: Impartiality, fairness, and justice for all people in social policy. Social equity accounts for systemic inequalities to ensure everyone in a community has access to the same opportunities and outcomes (NASEM, 2022d).

EXPOSURE: The situation of people, infrastructure, housing, production capacities, and other tangible human assets located in hazard-prone areas (UNDRR, n.d.-b).

EXTREME EVENT, CLIMATE-RELATED: Climate-related extreme events either persist longer than weather events or emerge from the accumulation of weather or climate events that persist over a longer period of time (USDA, 2024). Over the past decade, research has demonstrated that climate change due to global warming has made many extreme weather events more likely, more intense, longer-lasting, or larger in scale than they would have been without it (NOAA, 2024).

EXTREME EVENT, WEATHER-RELATED: An event that is rare at a particular place and time of year. Definitions of "rare" vary, but an extreme weather event would normally be as rare as or rarer than the event with a 10% probability of exceedance based on past observations. By definition, the characteristics of what is called extreme weather may vary from place to place in an absolute sense (IPCC, 2022). Weather-related extreme events are often short-lived and include heat waves, freezes, heavy downpours, tornadoes, tropical cyclones, and floods (USDA, 2024).

HAZARD: The potential occurrence of a natural or human-induced physical event or trend that may cause loss of life, injury, or other health impacts, as well as damage and loss to property, infrastructure, livelihoods, service provision, ecosystems, and environmental resources (USGCRP, 2023).

HAZARD MITIGATION: Steps taken before an event to reduce the exposure of people and property to environmental hazards and to reduce the negative impacts of those hazards (NRC, 2012).

IMPACTS (DISASTER): The consequences of realized risks on natural and human systems, where risks result from the interactions of climate-related hazards (including extreme weather/climate events), exposure, and vulnerability. Impacts generally refer to effects on lives, livelihoods, health, and well-being; ecosystems and species; economic, social, and cultural assets; services; and infrastructure (adapted from USGCRP, 2023 definition of Impacts (climate)).

INFRASTRUCTURE: Physical networks (systems and facilities) that provide functions and services to the community. Infrastructure systems include transportation, energy, communications, water, and wastewater systems. Building clusters (buildings with common functions) and supporting infrastructure systems are organized by functional categories, such as health, economy, education, or housing, for planning purposes (NIST, 2020).

PARTICIPATORY PLANNING: Participatory planning processes seek to engage and empower community members, especially marginalized people and those disproportionately affected by disasters to proactively engage in planning processes that build close partnerships with local and regional governments and strengthen community leadership and local capacities, including by providing flexible resources—such as financing—directly to community-based organizations to help them prepare and plan for, absorb, recover from, and more successfully adapt to adverse events (modified and adapted from Mfitumukiza et al., 2020).

RECOVERY: Those capabilities necessary to assist communities affected by an incident to recover effectively, including, but not limited to, rebuilding infrastructure systems; providing adequate interim and long-term housing for survivors; restoring health, social, and community services; promoting economic development; and restoring natural and cultural resources (FEMA, n.d., NDRF - National Disaster Recovery Framework).

RESILIENCE: The ability to prepare and plan for, absorb, recover from, and more successfully adapt to adverse events (NRC, 2012).

RESILIENCE, COMMUNITY: The ability of communities to withstand and recover and learn from past cumulative or compounding disasters to strengthen future response and recovery efforts. This can include, but is not limited to, physical and psychological health of the population, social and economic equity and well-being of the community, effective risk communication, integration of organizations (governmental and nongovernmental) in planning, response, and recovery (USGCRP, 2023).

SOCIAL CAPITAL: Features of social organization, such as networks, norms, and trust, that facilitate coordination and cooperation for mutual benefit (Putnam, 1993).

SOCIAL DETERMINANTS OF HEALTH: The conditions in the environments where people are born, live, learn, work, play, worship, and age that affect a wide range of health, functioning, and quality-of-life outcomes and risks. Social determinants of health can be grouped into five domains: economic stability, education access and quality, health care access and quality, neighborhood and built environment, and social and community context (USGCRP, 2023).

STRUCTURAL RACISM: The public and private policies, institutional practices, norms, and cultural representations that inherently procure unequal freedom, opportunity, value, resources, advantage, restrictions, constraints, or disadvantage for individuals and populations according to their race and ethnicity both across the life course and between generations (NASEM, 2022c).

TRANSFORMATIVE ADAPTATION: Adaptation that changes the fundamental attributes of a social–ecological system, often involving persistent, novel, and significant changes to institutions, behaviors, values, and/or technology in anticipation of climate change and its impacts (USGCRP, 2023).

UNDERSERVED COMMUNITY: A community with environmental justice concerns and/or vulnerable populations, including people of color, low income, rural, tribal, indigenous, and homeless populations (U.S. EPA, 2019), including small communities (fewer than 1,000 people in population) that lack the resources to carry out resilience and public health building efforts (NASEM, 2023c).

VULNERABILITY: The propensity or predisposition to be adversely affected. Vulnerability encompasses a variety of concepts and elements including sensitivity or susceptibility to harm and lack of capacity to cope and adapt (IPCC, 2021).

VULNERABLE COMMUNITY: Communities of people who face disproportionate and unequal risks from projected and realized climate change impacts and who are least able to anticipate, cope with, and recover from these adverse impacts. Socioeconomic factors may include, but are not limited to, income, educational attainment, race and ethnicity, and age (USGCRP, 2023).

WINTER STORM URI: The unofficial name of a major North American winter storm that affected large portions of Texas and Louisiana during the weeks of February 2021.

Appendix B

Public Session Meeting Agendas

CONSENSUS STUDY
COMPOUNDING DISASTERS IN GULF COAST COMMUNITIES, 2020–2021:
IMPACTS, FINDINGS, AND LESSONS LEARNED

PUBLIC AGENDA

October 10–11, 2022

University of Houston
Bayou City Room
University of Houston Student Center South

Land Acknowledgment
We would like to acknowledge that the land we occupy today has served as a site of Indigenous peoples, specifically the Atakapa-Ishak, Tāp Pīlam Coahuiltecan, the Sana band of the Tonkawa tribe, and Karankawa nations.

Meeting Objectives
- Learn from the experiences of public officials and non-governmental leaders.
- Identify and discuss lessons-learned and factors that enabled or could enable communities to successfully plan for, respond to, recover from, and mitigate the impacts of multiple, compounding disasters.

DAY 1 - OPEN SESSION – Welcome

8:30 – 9:00   Coffee and Light Breakfast Refreshments
                   Astrodome Room, University of Houston Student
                   Center South

**DAY 1 - OPEN SESSION – Session 1 – Non-Governmental Leaders –
Harris County**

9:00 – 9:05   Opening Remarks and Meeting Objectives
                   Roy Wright, *Committee Chair*
                   President and CEO
                   Insurance Institute for Business & Home Safety
                   (IBHS)

9:05 – 9:10   Overview of Facilitated Discussion Engagement Protocols
                   Tracy Kijewski-Correa, Ph.D., *Professor, University of
                   Notre Dame*

9:10 – 10:40   Facilitated Discussion
                   Craig Colten, Ph.D., *Professor Emeritus, Louisiana
                   State University*
                   Tracy Kijewski-Correa, Ph.D., *Professor, University of
                   Notre Dame*

                   Participants:
                   Charles K. Blake, *CEO, American Red Cross, Texas
                   Gulf Coast Region*
                   Doris Brown, *Co-Director of Community Research,
                   Organizing, and Special Events, West Street Recovery*
                   Ken Farris, *Director, Houston Rebuild*
                   Bob Harvey, *President and CEO, Greater Houston
                   Partnership*
                   Lisa Spivey, *Director of Preparedness Operations,
                   SouthEast Texas Regional Advisory Council*

10:40 – 11:00   Questions from the Study Committee Members

11:00   **Adjourn**

DAY 1 - OPEN SESSION – Session 2 — Public Officials — Harris County

11:15 – 11:20 Opening Remarks and Meeting Objectives
Roy Wright, *Committee Chair*
President and CEO
Insurance Institute for Business & Home Safety (IBHS)

11:20 – 11:25 Overview of Facilitated Discussion Engagement Protocols
Tracy Kijewski-Correa, Ph.D., *Professor, University of Notre Dame*

11:25 – 12:55 Facilitated Discussion
Craig Colten, Ph.D., *Professor Emeritus, Louisiana State University*
Tracy Kijewski-Correa, Ph.D., *Professor, University of Notre Dame*

Participants:
John Fleming, *Houston Health Department*
Darrell Hahn, *Interim Permits Manager, Harris County Engineering Department*
Thomas Munoz, *Director, City of Houston Department of Energy*
Barbie Robinson, *Executive Director, Harris County Public Health*

12:55 – 1:15 Questions from the Study Committee Members

1:15 **Adjourn**

LUNCH

1:15 – 2:00 Focus group participants are invited to lunch in the Astrodome Room, University of Houston Student Union

DAY 1 - OPEN SESSION – Session 3 – Federal and State Officials

2:00 – 2:05   Opening Remarks and Meeting Objectives
Roy Wright, *Committee Chair*
President and CEO
Insurance Institute for Business & Home Safety
(IBHS)

2:05 – 2:10   Overview of Facilitated Discussion Engagement Protocols
Tracy Kijewski-Correa, Ph.D., *Professor, University of Notre Dame*

2:10 – 3:40   Facilitated Discussion
Craig Colten, Ph.D., *Professor Emeritus, Louisiana State University*
Tracy Kijewski-Correa, Ph.D., *Professor, University of Notre Dame*

Participants:
Jet Hays, *Deputy Director: Integration, Texas General Land Office*
Francisco Sanchez, *Associate Administrator, SBA*
Benjamin Abbott, *Federal Coordinating Officer, US DHS*

3:40 – 4:00   Questions from the Committee Members

4:00   **Adjourn**

***

DAY 2 - OPEN SESSION – Welcome
8:30 – 9:00   Coffee and Light Breakfast Refreshments

**DAY 2 - OPEN SESSION – Session 4 – Non-Governmental Leaders –
Galveston County**

9:00 – 9:05   Opening Remarks and Meeting Objectives
          Roy Wright, *Committee Chair*
          President and CEO
             Insurance Institute for Business & Home Safety
             (IBHS)

9:05 – 9:10   Overview of Facilitated Discussion Engagement Protocols
          Tracy Kijewski-Correa, Ph.D., *Professor, University of
          Notre Dame*

9:10 – 10:40   Facilitated Discussion
          Craig Colten, Ph.D., *Professor Emeritus, Louisiana
          State University*
          Tracy Kijewski-Correa, Ph.D., *Professor, University of
          Notre Dame*

          Participants:
          Charles K. Blake, *CEO, American Red Cross, Texas
          Gulf Coast Region*
          Jessica Debalski, *American Red Cross*
          John Herrmann, *Coordinator, Galveston County
          Community Emergency Response Team*
          Lynda Perez, *Galveston County Recovers*
          Leslie Mamud, *Galveston County Recovers*

10:40 – 11:00 Questions from the Committee Members

11:00   **Adjourn**

**OPEN SESSION – Session 5 – Governmental Officials – Galveston
County**

11:15 – 11:20 Opening Remarks and Meeting Objectives
          Roy Wright, *Committee Chair*
          President and CEO
             Insurance Institute for Business & Home Safety
             (IBHS)

11:20 – 11:25   Overview of Facilitated Discussion Engagement Protocols
Tracy Kijewski-Correa, Ph.D., *Professor, University of Notre Dame*

11:25 – 12:55   Facilitated Discussion
Craig Colten, Ph.D., *Professor Emeritus, Louisiana State University*
Tracy Kijewski-Correa, Ph.D., *Professor, University of Notre Dame*

Participants:
Renaye Ochoa, *Lieutenant, City of Galveston*
Mona Purgason, *President, Galveston Housing Authority*
Joe Tumbleson, *Texas City Emergency Manager*

12:55 – 1:15   Questions from the Committee Members

1:15   **Adjourn**

LUNCH
1:15   Speakers are invited to take a box lunch from the South Heights Room, University of Houston Student Union

December 7–8, 2022

Lake Charles, LA

Land Acknowledgment
We would like to acknowledge that this space where we convene was once occupied by Indigenous peoples, specifically the Atakapa-Ishak Nation, a Nation that remains extant in the region today.

Meeting Objectives
- Learn from the experiences of public officials and non-governmental leaders.
- Identify and discuss lessons-learned and factors that enabled or could enable communities to successfully plan for, respond to, recover from, and mitigate the impacts of multiple, compounding disasters.

AGENDA

DAY 1 - OPEN SESSION – Welcome
8:30 – 9:00   Coffee and Light Breakfast Refreshments
            Bayou I Room, Holiday Inn and Suites Lake Charles South

DAY 1 - OPEN SESSION – Session 1 – Non-Governmental Leaders – SWLA
9:00 – 9:05   Opening Remarks and Meeting Objectives
            Roy Wright, *Committee Chair*
            President and CEO
            Insurance Institute for Business & Home Safety (IBHS)

9:05 – 9:10   Overview of Facilitated Discussion Engagement Protocols
            Tracy Kijewski-Correa, Ph.D., *Professor, University of Notre Dame*

9:10 – 10:40   Facilitated Discussion
    Craig Colten, Ph.D., *Professor Emeritus, Louisiana State University*
    Tracy Kijewski-Correa, Ph.D., *Professor, University of Notre Dame*

    Key Informants/Participants:
    Denise Durel, *President and CEO, United Way of SWLA*
    Jacqueline Green, *Executive Director, Calcasieu Council on Aging*
    Traci Hedrick, *Community Services Director, Imperial Calcasieu Human Services Authority*
    Angie Herr, *Director of Programs, Calcasieu Council on Aging*
    John O'Donnell, *Executive Director SWLA Region, SBP*
    Davelyn Patrick, *Community Resource Specialist, Families Helping Families*
    Keyon Payton, *Social Justice Committee Chair, National Baptist Convention of America International*
    Chele Rider, *Regional CEO, American Red Cross Louisiana Region*
    Jim Rock*, Vice Chair, Community Foundation SWLA*
    Jeffrey Starkovich, *Pastor, St. Pius X Catholic Church*

10:40 – 11:00   Questions from the Study Committee Members

11:00   **Adjourn**

LUNCH
11:00 – 12:30   Focus group participants are invited to a catered lunch in the Bayou I Room

DAY 1 - OPEN SESSION – Session 2 — Public Officials — Cameron Parish
12:30 – 12:40   Opening Remarks and Meeting Objectives
    Roy Wright, *Committee Chair*
    President and CEO
    Insurance Institute for Business & Home Safety (IBHS)

12:40 – 12:45   Overview of Facilitated Discussion Engagement Protocols
                 Tracy Kijewski-Correa, Ph.D., *Professor, University of Notre Dame*

12:45 – 2:15   Facilitated Discussion
                 Craig Colten, Ph.D., *Professor Emeritus, Louisiana State University*
                 Tracy Kijewski-Correa, Ph.D., *Professor, University of Notre Dame*

                 Key Informants/Participants:
                 Katie Armentor, *Cameron Parish Administrator, Cameron Parish Police Jury*
                 Danny Lavergne, *Director, Cameron Parish Office of Homeland Security and Emergency Preparedness*
                 Braden Ryder, *Chief Operating Officer, South Cameron Memorial Hospital*
                 Scott Trahan, *District 5 Representative, Cameron Parish Police Jury*

2:15 – 2:30   Questions from the Study Committee Members

2:30   **Adjourn**

***

DAY 2 - OPEN SESSION – Welcome
9:00 – 9:30   Coffee and Light Breakfast Refreshments
                 Bayou I Room, Holiday Inn and Suites Lake Charles South

DAY 2 - OPEN SESSION – Session 4 – Public Officials—Calcasieu Parish
9:30 – 9:35   Opening Remarks and Meeting Objectives
                 Roy Wright, *Committee Chair*
                 President and CEO
                 Insurance Institute for Business & Home Safety (IBHS)

9:35 – 9:40   Overview of Facilitated Discussion Engagement Protocols
Tracy Kijewski-Correa, Ph.D., *Professor, University of Notre Dame*

9:40 – 11:10   Facilitated Discussion
Craig Colten, Ph.D., *Professor Emeritus, Louisiana State University*
Tracy Kijewski-Correa, Ph.D., *Professor, University of Notre Dame*

Key Informants/Participants:
Bryan Beam, *Calcasieu Parish Administrator*
Michael Carter, *Emergency Operations Director, Imperial Calcasieu Human Services Authority*
Erika Doshier, *Executive Assistant, Calcasieu Parish Human Services*
Erika Fontenot, *Operations Manager, Calcasieu Parish Office of Homeland Security and Emergency Preparedness*
Nic Hunter, *Mayor, City of Lake Charles*
Tanya McGee, *Executive Director, Imperial Calcasieu Human Services Authority*
Nicole Miller, *Calcasieu Parish Police Jury Program Manager and Disaster Housing Recovery Board Chair*
Teri Talbot, *Accountant, Calcasieu Parish Public Health Services*
Ben Taylor, *Director, Lake Charles Housing Authority*

11:10 – 11:30 Questions from the Committee Members

11:30   **Adjourn**

LUNCH
11:30 - 12:30   Focus group participants are invited to a catered lunch in the conference room

OPEN SESSION – Session 5 – State, Federal, and Tribal Officials
12:30 – 12:40   Opening Remarks and Meeting Objectives
                Roy Wright, *Committee Chair*
                President and CEO
                    Insurance Institute for Business & Home Safety
                    (IBHS)

12:40 – 12:45   Overview of Facilitated Discussion Engagement Protocols
                Tracy Kijewski-Correa, Ph.D., *Professor, University of
                Notre Dame*

12:45 – 2:15   Facilitated Discussion
                Craig Colten, Ph.D., *Professor Emeritus, Louisiana
                    State University*
                Tracy Kijewski-Correa, Ph.D., *Professor, University of
                Notre Dame*

                Key Informants/Participants:
                Benjamin Abbott, *Federal Coordinating Officer, FEMA
                    Region 6*
                Lacey Cavanaugh, *Region 5 Director, Louisiana
                    Department of Health*
                Pat Forbes, *Executive Director, Louisiana Office of
                    Community Development*
                Chris Hector, *Administrative Director, Louisiana
                    Emergency Response Network*
                Tanisha Hubbard, *Program Coordinator, Louisiana
                    Department of Health, Office of Behavioral Health,
                    Louisiana Spirit Counselling Program*
                Raymond Rodriguez, *Disaster Response and Recovery
                    Director, Louisiana Housing Corporation*
                Casey Tingle, *Acting Director, Governor's Office of
                    Homeland Security and Emergency Preparedness*

2:15 – 2:30   Questions from the Committee Members

2:30   **Adjourn**

December 12, 2022

Housing and Displacement Virtual Panel

Meeting Objectives
- Learn from the experiences of practitioners and subject matter experts in the areas of housing and displacement based in case-study locations.
- Identify and discuss lessons-learned and factors that enabled or could enable communities to successfully plan for, respond to, recover from, and mitigate the impacts of multiple, compounding disasters.

AGENDA

OPEN SESSION – Housing and Displacement Panel

1:00 – 1:05 Opening Remarks and Meeting Objectives
 Roy Wright, *Committee Chair*
 President and CEO
  Insurance Institute for Business & Home Safety (IBHS)

1:05 – 1:10 Overview of Discussion Topics and Format
 Craig Colten, Ph.D., *Professor Emeritus, Louisiana State University*

1:10 – 2:10 Facilitated Discussion
 Craig Colten, Ph.D., *Professor Emeritus, Louisiana State University*

 Key Informants/Participants:
 Arthur Johnson, *Chief Executive Director, Lower 9th Ward Center for Sustainable Engagement and Development*
 Natalie Manning, *Communications Mentor, Lower 9th Ward Center for Sustainable Engagement and Development*
 Madison Sloan, *Director of Disaster Recover and Fair Housing Project, Texas Appleseed*

Shannon Van Zandt, *Professor of Landscape Architecture and Urban Planning, Texas A&M University*

2:10 – 2:30   Questions from the Study Committee Members

2:30   **Adjourn**

January 9, 2023

Meteorological and Oceanographic Influences on
Storm Intensity Virtual Panel

Meeting Objectives
- Learn from subject matter experts in the areas of meteorology and oceanography about factors that can influence storm intensity and frequency.

AGENDA

OPEN SESSION – Meteorological and Oceanographic Influences on Storm Intensity Panel

10:00 – 10:05   Opening Remarks and Meeting Objectives
Roy Wright, *Committee Chair*
President and CEO
Insurance Institute for Business & Home Safety
(IBHS)

10:05 – 10:10   Overview of Presentation Topics and Format
J. Marshall Shepherd, Ph.D., *Professor, University of Georgia*

10:10 – 10:30   Presentation: Aspects of Rapid Intensification
Bill Read, *Director (Retired), U.S. National Hurricane Center*

10:30 – 10:50   Presentation: Compounding Disasters: Rapid Intensification (RI)
Kieran Bhatia, *Vice President, Climate Change Perils Advisory at Guy Carpenter*

10:50 – 11:30   Questions from the Study Committee Members

11:30   **Adjourn**

January 31 – February 1, 2023

University of South Alabama
Mobile, AL

Land Acknowledgment
We would like to acknowledge that we are currently occupying the lands once inhabited by the Chahta Yakni Nation, commonly known as the Choctaw Nation, and Pensacola Nation.

Meeting Objectives
- Learn from the experiences of public officials and non-governmental leaders.
- Identify and discuss lessons-learned and factors that enabled or could enable communities to successfully plan for, respond to, recover from, and mitigate the impacts of multiple, compounding disasters.

AGENDA

DAY 1 - OPEN SESSION – Welcome
8:30 – 9:00   Coffee and Light Breakfast Refreshments
                          John Counts Room, Mitchell Center, University of South Alabama

DAY 1 - OPEN SESSION – Session 1 – Non-Governmental Leaders
9:00 – 9:05   Opening Remarks and Meeting Objectives
                          Roy Wright, *Committee Chair*
                          President and CEO
                              Insurance Institute for Business & Home Safety (IBHS)

9:05 – 9:10   Overview of Facilitated Discussion Engagement Protocols
                          Tracy Kijewski-Correa, Ph.D., *Professor, University of Notre Dame*

9:10 – 10:40   Facilitated Discussion
                          Jeff Byard, *Vice President of Operations, Team Rubicon*
                          Tracy Kijewski-Correa, Ph.D., *Professor, University of Notre Dame*

Key Informants/Participants:

Martha Arietta, *Interim Director, Center for Healthy Communities*

Chandra Brown Stewart, *Executive Director, Lifelines Counseling*

Jill Chenoweth, *President and CEO, United Way of Southwest Alabama*

Janel Lowman, *Senior Community Outreach Manager, Mitchell Cancer Institute*

Lana Mummah, *Finance and Program Director, United Way of Baldwin County*

Laura Myers, *Director, Alabama Center for Insurance Information and Research*

Danny Patterson, *Coalition Coordinator, Gulf States Health Policy Center*

Julie Shiyou-Woodard, *President and CEO, Smart Home America*

Trista Stout-Walker, *Vice President of Community Impact, United Way of Southwest Alabama*

Kim Lien Tran, *Community Health Worker, Boat People SOS*

10:40 – 11:00   Questions from the Study Committee Members

11:00   **Adjourn**

LUNCH
11:00 – 12:30   Focus group participants are invited to a catered lunch in the John Counts Room

DAY 1 - OPEN SESSION – Session 2 — Public Officials — Baldwin County
12:30 – 12:40   Opening Remarks and Meeting Objectives
Roy Wright, *Committee Chair*
President and CEO
Insurance Institute for Business & Home Safety
(IBHS)

12:40 – 12:45   Overview of Facilitated Discussion Engagement Protocols
      Tracy Kijewski-Correa, Ph.D. *Professor, University of Notre Dame*

12:45 – 2:15   Facilitated Discussion
      Jeff Byard, *Vice President of Operations, Team Rubicon*
      Tracy Kijewski-Correa, Ph.D., *Professor, University of Notre Dame*

      Key Informants/Participants:
      Dan Bond, *Environmental/Grants Coordinator, City of Gulf Shores*
      Brandan Franklin, *Chief Building Official, Gulf Shores Building Department*
      Ronnie Huskey, *Public Works Director, City of Daphne*

2:15 – 2:30   Questions from the Study Committee Members

2:30   **Adjourn**

***

DAY 2 - OPEN SESSION – Welcome
8:30 – 9:00   Coffee and Light Breakfast Refreshments
      John Counts Room, Mitchell Center, University of South Alabama

DAY 2 - OPEN SESSION – Session 3– Public Officials—Mobile County
9:00 – 9:05   Opening Remarks and Meeting Objectives
      Roy Wright, *Committee Chair*
      President and CEO
        Insurance Institute for Business & Home Safety (IBHS)

9:05 – 9:10   Overview of Facilitated Discussion Engagement Protocols
      Tracy Kijewski-Correa, Ph.D., *Professor, University of Notre Dame*

9:10 – 10:40   Facilitated Discussion
> **Jeff Byard,** *Vice President of Operations, Team Rubicon*
> **Tracy Kijewski-Correa, Ph.D.,** *Professor, University of Notre Dame*

> Key Informants/Participants:
> **Doug Cooper,** *Plans and Operations Officer, Mobile County Emergency Management Agency*
> **Casi Callaway,** *Chief Resilience Officer, City of Mobile*
> **Erin Coker,** *Emergency Preparedness Administrator, Mobile County Health Department*
> **Chuck Harben,** *Supervisor of Accounts Payable and Risk Management, Mobile County Public Schools*
> **Kevin Michaels,** *Mobile County Health Officer, Mobile County Health Department*
> **Beverly Reed,** *Assistant Director of Community and Housing Development, Neighborhood Development City of Mobile*
> **Ray Richardson,** *Environmental Manager, Mobile Department of Environmental Services*
> **Tina Sanchez,** *Director, Mobile County Environmental Services*
> **Stephanie Woods-Crawford,** *Executive Director, Mobile County Health Department*

10:40 – 11:00 Questions from the Committee Members

11:00   **Adjourn**

LUNCH
11:00 - 12:30 Focus group participants are invited to a catered lunch in the John Counts Room

OPEN SESSION – Session 4 – State and Federal Officials
12:30 – 12:40 Opening Remarks and Meeting Objectives
> **Roy Wright**, *Committee Chair*
> President and CEO
> > Insurance Institute for Business & Home Safety (IBHS)

12:40 – 12:45   Overview of Facilitated Discussion Engagement Protocols
                  Tracy Kijewski-Correa, Ph.D., *Professor, University of Notre Dame*

12:45 – 2:15   Facilitated Discussion
                  Jeff Byard, *Vice President of Operations, Team Rubicon*
                  Tracy Kijewski-Correa, Ph.D., *Professor, University of Notre Dame*

                  Key Informants/Participants:
                  Jamey Durham, *Director of the Bureau of Prevention, Promotion and Support, Alabama Department of Public Health*
                  Kenneth Free, *Director, HUD Alabama Field Office*
                  Allan Jarvis, *Federal Coordinating Officer, FEMA Region 4*
                  Brian Powell, *Director, Alabama Department of Insurance*
                  David Tidwell, *Geologist and GIS Specialist, Geological Survey of Alabama*

2:15 – 2:30   Questions from the Committee Members

2:30   **Adjourn**

February 8, 2023

Chief Resilience Officers Virtual Panel

Meeting Objectives
- Learn from the experiences of Chief Resilience Officers based in Gulf Coast locations.
- Identify and discuss lessons-learned and factors that enabled or could enable communities to successfully plan for, respond to, recover from, and mitigate the impacts of multiple, compounding disasters.

AGENDA

OPEN SESSION – Chief Resilience Officers Panel

10:00 – 10:05   Opening Remarks and Meeting Objectives
Roy Wright, *Committee Chair*
President and CEO
Insurance Institute for Business & Home Safety
(IBHS)

10:05 – 1:10   Overview of Discussion Topics and Format
Roy Wright, *Committee Chair*
President and CEO
Insurance Institute for Business & Home Safety
(IBHS)

10:10 – 11:10   Facilitated Discussion
Roy Wright, *Committee Chair*
President and CEO
Insurance Institute for Business & Home Safety
(IBHS)

Key Informants/Participants:
Casi Callaway, *Chief Resilience Officer, City of Mobile*
Charles Sutcliffe, *Chief Resilience Officer, Louisiana Governor's Office of Coastal Activities*
Joseph Threat, *Chief Resilience Officer, City of New Orleans*

11:10 – 11:30   Questions from the Study Committee Members

11:30   **Adjourn**

April 27, 2023

Community Stakeholders Perspectives Virtual Panel

Meeting Objectives
- Learn from the experiences of community stakeholders who experienced the impacts of multiple successive and/or concurrent disasters first-hand.
- Identify and discuss lessons-learned and factors that enabled or could enable communities to successfully plan for, respond to, recover from, and mitigate the impacts of multiple, compounding disasters.

AGENDA

OPEN SESSION – Community Stakeholder Perspectives

2:00 – 2:05   Opening Remarks and Meeting Objectives
Roy Wright, *Committee Chair*
President and CEO
Insurance Institute for Business & Home Safety (IBHS)

2:05 – 2:10   Overview of Discussion Topics and Format
Chauncia Willis, *CEO, Institute for Diversity and Inclusion in Emergency Management*

2:10 – 3:10   Facilitated Discussion
Chauncia Willis, *CEO, Institute for Diversity and Inclusion in Emergency Management*

Key Informants/Participants:
Louise Billiot, *Tribal Official, United Houma Nation of Louisiana*
Lanor Curole, *Director and Administrator, United Houma Nation of Louisiana*
Deme J.R. Naquin, *Chief, Jean Charles Choctaw Nation*
Roishetta Sibley Ozane, *Founder, Director and CEO, the Vessel Project of Louisiana*

3:10 – 3:30   Questions from the Study Committee Members

3:30   **Adjourn**

Appendix C

Commissioned Paper Abstracts

Note: The commissioned papers may be accessed on the National Academies Press website https://nap.nationalacademies.org/catalog/27170

Compounding Disasters in Gulf Coast Communities 2020–2021: Impacts, Findings, and Lessons Learned in Jefferson Davis and Marion Counties, Mississippi

Jennifer Trivedi

Abstract:

This work explores disasters that affected Jefferson Davis and Marion Counties, Mississippi, in 2020 and 2021 as part of an effort to better understand compounding disasters in the Gulf Coast region. Understanding the ways these two counties and the people who live there were affected by, coped with, and moved through these disasters as concurrent processes, rather than static or momentary events (affected by what came before and what continued after), is important for understanding the larger implications, history, and future of compounding disasters in Gulf Coast communities. In the United States, since COVID-19 spread nationwide, other disasters have become compounding disasters as they have overlapped in time and space with COVID-19, leaving emergency managers, government officials, businesses, and residents to respond to COVID-19 and other disasters simultaneously—a situation complicated by local cultural, historical, political, and economic contexts.

How a Changing Societal Landscape Is Shaping Gulf Coast Tropical Cyclone and Tornado Disasters

Stephen Strader

Abstract:

This study examines how tropical cyclone and tornado risk, exposure, and vulnerability are changing over time within the conterminous United States, Gulf Coast region, and eight representative counties and parishes in the Gulf Coast region. Findings indicate that although landfalling tropical cyclone and tornado events have not increased in frequency in the Gulf Coast region, societal and built environment exposure to these events has been rapidly amplified during the last 80 years. Since 1940, the numbers of people and housing units in the Gulf Coast region have increased by 433 percent and 2,818 percent, respectively. Results outlining changes in social vulnerability metrics from 2000 to 2018 are varied for the conterminous United States, Gulf Coast region, and selected counties. For instance, socioeconomic vulnerability measures assessed in this study (e.g., income, poverty, education) increased during the 18-year vulnerability analysis period across all domains, while measures of vulnerability in household composition and disability, minority status and language, and housing type and transportation have slightly decreased or stayed the same over time in the study domains. Nevertheless, social vulnerability in the Gulf Coast region continues to be approximately 25 percent higher than in the conterminous United States as a whole. Outcomes from this study may be used to assist policymakers, emergency managers, and stakeholders in developing or improving existing hazard mitigation plans, including building community resilience to hazards in the context of a changing climate and society.

Appendix D

Committee Member and Staff Biographical Sketches

Roy E. Wright, M.P.A. *(Chair),* is president and CEO of the Institute for Business and Home Safety (IBHS). Mr. Wright joined IBHS in 2018 with more than 20 years of experience in insurance, risk management, mitigation, and resilience planning. Convinced that the continuing cycle of human suffering that strikes families and communities in the wake of severe weather can be broken, Mr. Wright leads a team of scientists and risk communicators who deliver strategies to build safer and stronger homes and businesses. IBHS's real-world impact enables the insurance industry and affected property owners to prevent avoidable losses. Mr. Wright joined IBHS from the Federal Emergency Management Agency (FEMA) where he served as the chief executive of the National Flood Insurance Program, led the agency's Federal Insurance and Mitigation Administration, and directed the resilience programs addressing earthquake, fire, flood, and wind risks. In these roles, he guided several programs that promote a risk-conscious culture, enable faster disaster recovery, and address long-term vulnerabilities to life, property, and well-being in communities across the United States. Prior to joining FEMA in 2007, he worked in public and private sector roles with Coray Gurnitz Strategy Consulting and the U.S. Department of the Interior. Mr. Wright earned a bachelor's degree in political science from Azusa Pacific University and a master of public administration from the George Washington University.

Jeff Byard is a Greyshirt and Marine Corps veteran serving as vice president of operations for Team Rubicon. In that capacity, he leads a highly functional and diverse team of staff and volunteers dedicated to delivering disaster relief services across the country. He develops strategies to grow connections with local, state, federal, and other partners to enhance effective and efficient service delivery. His résumé includes being the associate administrator, Office of Response and Recovery at FEMA, and executive operations officer at the Alabama Emergency Management Agency. Jeff honorably served in the United States Marine Corps from 1990 to 1994. Mr. Byard holds a bachelor of science degree from Troy University and is a graduate of the FEMA Executive Academy (Cohort V) and the Alabama Public Safety Leadership Academy (Cohort I). Mr. Byard is a charter member of the Region IV Regional Advisory Council. Mr. Byard received the Department of Homeland Security Outstanding Service Medal and the Distinguished Public Safety Award from the United States Coast Guard.

Craig E. Colten, Ph.D., is a professor emeritus of Louisiana State University. Before retirement in 2021, he was the Carl O. Sauer Professor of Geography and Anthropology. He held prior positions with the Illinois Department of Natural Resources, PHR Environmental Consultants, and Texas State University. He has published several books dealing with the hazards and society on the Gulf Coast, including the award-winning *An Unnatural Metropolis* (2005), along with *Perilous Place, Powerful Storms* (2009), *Southern Waters* (2014), and *State of Disaster* (2021). His research has been funded by the State of Illinois, the National Park Service, the Corps of Engineers, Oak Ridge National Laboratory, the National Institutes of Environment Health Science, the Water Institute of the Gulf, the National Academy of Sciences, and the Mellon Foundation. The results of his work have appeared in a range of academic journals. He has received a Rainmaker Award from Louisiana State University and is a fellow of the American Association of Geographers.

Tracy L. Kijewski-Correa, Ph.D., is the Leo E. and Patti Ruth Linbeck Collegiate Chair and associate professor in the Department of Civil and Environmental Engineering and Earth Sciences at the University of Notre Dame. Jointly appointed as an associate professor of global affairs, she also serves as co-director of the Keough School's Integration Lab (i-Lab) and faculty fellow at a number of institutes focused on global development and real estate. Her research is dedicated to enhancing the resilience and sustainability of hazard-exposed communities, with an emphasis on conceiving holistic responses to infrastructure vulnerabilities and developing tools that support science-informed decision-making by diverse stakeholders. She currently serves as the inaugural director of the National Science Foundation's Structural Extreme Event Reconnaissance (StEER) Network, coordinating damage assessments following hazard events worldwide. Her contributions have been recognized by awards from the American Society of Civil Engineering, American Political Science Association, Institution of Civil Engineers, International Association for Wind Engineering, and American Association for Wind Engineering. Dr. Kijewski-Correa is formally trained as a civil engineer with a specialization in structural engineering, earning her bachelor of science, master of science, and Ph.D. from the University of Notre Dame.

J. Marshall Shepherd, Ph.D., is a leading international expert in weather and climate and is the Georgia Athletic Association Distinguished Professor of Geography and Atmospheric Sciences at the University of Georgia (UGA). Dr. Shepherd was the 2013 president of the American Meteorological Society (AMS), the nation's largest and oldest professional/science society in the atmospheric and related sciences. Dr. Shepherd serves as director of the University of Georgia's Atmospheric Sciences Program and full professor in the Department of Geography where he was a previous associate department head. He is also the host of the Weather Channel's award-winning show *Weather Geeks*, a pioneering Sunday talk podcast/show, and a contributor to *Forbes Magazine*. In 2021, Dr. Shepherd was elected to the National Academy of Sciences, National Academy of Engineering, and the American Academy of Arts and Sciences. He is the only faculty member in UGA history to be elected to all three, which are considered some of the highest honors bestowed on a scientist or engineer. The same year, Dr. Shepherd also received the Friends of the Planet Award from the National Council of Science Educators and the American Geological Institutes Award for Engagement in the Geosciences. Dr. Shepherd received the 2020 Mani L. Bhaumik Award for Public Engagement with Science from the American Association for the Advancement of Science, the 2019 AGU Climate Communication Prize, and the 2018 prestigious AMS Helmut Landsberg Award for pioneering and significant work in urban climate. In 2017, he was honored with the AMS Brooks Award, a high honor within the field of meteorology. Ted Turner and his Captain

Planet Foundation honored Dr. Shepherd in 2014 with its Protector of the Earth Award. Prior recipients include Erin Brockovich and former Environmental Protection Agency (EPA) administrator Lisa Jackson. He is also the 2015 recipient of the Association of American Geographers (AAG) Media Achievement Award, the Florida State University Grads Made Good Award, and the UGA Franklin College of Arts and Sciences Sandy Beaver Award for Excellence in Teaching. In 2015, he was invited to moderate the White House Champions for Change event. He is an alumnus of the prestigious SEC Academic Leadership Fellows Program. Prior to UGA, Dr. Shepherd spent 12 years as a research meteorologist at NASA-Goddard Space Flight Center and was deputy project scientist for the Global Precipitation Measurement mission, a multinational space mission that launched in 2014. President Bush honored him on May 4, 2004, at the White House with the Presidential Early Career Award for pioneering scientific research in weather and climate science. Dr. Shepherd is a fellow of the American Meteorological Society. Two national magazines, the AMS, and Florida State University have also recognized him for his significant contributions. He was the 2016 Spring Undergraduate Commencement speaker at his three-time alma mater, Florida State University. He was also the 2017 Graduate Commencement speaker at the University of Georgia. Dr. Shepherd received his B.S., M.S., and Ph.D. in physical meteorology from Florida State University. He was the first African American to receive a Ph.D. from the Florida State University Department of Meteorology, one of the nation's oldest and respected. He is also the second African American to preside over the American Meteorological Society.

James Shultz, Ph.D., is a population health scientist and associate professor in the Department of Public Health Sciences, University of Miami Miller School of Medicine, where he holds multiple leadership roles. First, Dr. Shultz is the director of the Center for Disaster and Extreme Event Preparedness (DEEP Center). Second, he is the director of P3H: Protect & Promote Population Health in Complex Crises. Third, he is the disaster public health lead for the Global Institute for Community Health and Development. Dr. Shultz teaches core public health courses in both residential and online formats and a disaster and emergency public health elective in the Graduate Programs in Public Health, University of Miami Miller School of Medicine. Dr. Shultz publishes and conducts research on themes of population health science, disaster behavioral health, protecting medically high-risk patients from disasters, climate change impacts on population health, complex and compounding disaster risks and resilience, global mental health, and structural violence. He is lead author on a popular textbook, *Public Health: An Introduction to the Science and Practice of Population Health*, second edition, along with Drs. Lisa Sullivan and Sandro Galea. Dr. Shultz holds a Ph.D. in behavioral epidemiology and a master of science degree in health behavior research from the University of Minnesota Division of Epidemiology and Community Health.

Chauncia T. Willis, M.P.A., is the co-founder and CEO of the Institute for Diversity and Inclusion in Emergency Management (I-DIEM). Ms. Willis is certified as an emergency manager, professional coach, and cultural diversity professional with more than 20 years of experience. Her expertise includes disaster management; national security event planning; leadership coaching; immigrant and refugee outreach; and diversity, equity, and inclusion training. Prior to co-founding I-DIEM, Ms. Willis served as the emergency manager in Tampa, Florida, for more than 14 years, where she developed successful programs benefiting marginalized populations. She has led national emergency planning efforts for political conventions, national football league games, and international award shows. In her role as CEO for I-DIEM, she leads the effort to integrate equity into all facets of disaster policy, programs, and practice with the goal of increasing cultural competence and mitigating the harmful effects of bias on underserved groups. Regarded as a national expert, she has provided witness testimony to the U.S. House of Representatives on multiple occasions on matters of diversity, equity, and inclusion, as well as policy implementation. Ms. Willis hails from St. Petersburg, Florida, and is a graduate of Loyola University New Orleans and Georgia State University, Andrew Young School of Policy Studies.

STAFF

Dan Burger, M.P.A., is a senior program manager for the Gulf Research Program (GRP) Division and leads the work of the Gulf Health and Resilience Board (GHRB). Mr. Burger's work concentrates on the intersection of climate change, human health, and disaster resilience and includes a broad portfolio of applied research, strategic partnerships, and capacity building. Prior to joining the National Academies of Sciences, Engineering, and Medicine, Mr. Burger held senior management and leadership positions for the South Carolina Department of Health and Environmental Control's

Coastal Zone Management Program focusing on policy and intergovernmental coastal resource management, hazard mitigation, and resilience planning efforts. Mr. Burger is also a founder and former chair of the Charleston Resilience Network, a collaborative effort among public, private, academic, and nongovernmental organizations to enhance regional decision-making and improve the resilience of social, physical, and economic systems. Mr. Burger has served on numerous regional and national advisory boards including an appointment to the National Academies' Resilient America Roundtable. Prior to his work in South Carolina, he worked to advance environmental public policy and build the capacity of nonprofit organizations and associations in Maryland. He is an honors graduate of Western Maryland (McDaniel) College and holds a master of public administration in urban public affairs from the College of Charleston.

Jessica Simms, Ph.D., is a program officer for the Health and Resilience Board in the Gulf Research Program. Her work and research lie at the intersection of climate, health, and equity and include community-based relocations, solastalgia, transdisciplinary approaches to solving complex challenges, community capacity-building for equitable access to resources, and compounding and cascading disasters. Dr. Simms has a B.A. from the University of California, Santa Cruz, in politics, an M.A. in geography from San Diego State University, and a Ph.D. in geography from Louisiana State University. For her dissertation research, she interviewed more than 100 coastal Louisiana residents who have already or are currently facing possible relocation decisions or displacement in order to understand the links among the influence and mobility of three factors: social relations (faith-based networks, civic organizations, family, cultural and heritage identities, etc.), inherent resilient practices, and place, including sense of and attachment to it. Before joining the GRP in October 2021, Dr. Simms worked for 4 years as the lead outreach and engagement coordinator for the Resettlement of Isle de Jean Charles in Terrebonne Parish, Louisiana.

Jennifer A. Cohen, M.P.H., is a senior program officer in the Gulf Research Program Division of the National Academies of Sciences, Engineering, and Medicine. Ms. Cohen has worked on a number of projects at the National Academies, including Organ Procurement and Transplantation, Clearing the Air: Asthma and Indoor Air Exposures, Veterans and Agent Orange: Update 2000 and Update 2014, Post-Vietnam Exposure in Agent Orange-Contaminated C-123 Aircraft, Health Effects of

Cannabis and Cannabinoids, Getting to Zero Alcohol-Impaired Driving Fatalities, and Social Isolation and Loneliness in Older Adults: Opportunities for the Health Care System. She was also the rapporteur for the workshop summary Challenges and Successes in Reducing Health Disparities. She received her undergraduate degree and M.P.H. from the University of Maryland, College Park.

Robert Gasior, M.P.H., works on improving health and community resilience to climate change and other disasters as a program officer at the Gulf Research Program of the National Academies of Sciences, Engineering, and Medicine. Previously, he directed several programs related to science and technology cooperation and science diplomacy. He directed the Partnerships for Enhanced Engagement in Research (PEER)/Liberia program, a $6 million initiative, which aimed to build subspecialty medical education capacity in Liberia to benefit the populations at large and specifically Ebola survivors. Additionally, he managed a portfolio of more than 40 research grants focused on maternal and child health, infectious disease, WASH (water, sanitation, and hygiene), tobacco control, renewable energy, and biodiversity located in more than 20 countries. He started his career at the National Academies with the Institute of Medicine's (IOM) Forum on Microbial Threats. Prior to the National Academies, he led service-learning and student leadership development initiatives at the Gerhart Center for Philanthropy and Civic Engagement at the American University in Cairo, Egypt. Mr. Gasior received a B.S. in biology and anthropology-zoology from the University of Michigan, Ann Arbor, and an M.P.H. degree in epidemiology at George Washington University.

Sasha Allison is a research associate in the Gulf Health and Resilience Board of the Gulf Research Program. She has been working with the GHRB since June 2022. Prior to joining the GRP, Ms. Allison earned her bachelor's degree in sociology with a minor in anthropology from Vassar College in Poughkeepsie, New York, in May 2022. Much of her previous undergraduate coursework and experience has shaped and solidified her interest in environmental and social policy, from summers spent researching land use policy in Central America to time as a political canvasser for housing and energy justice to work as an instructor for a summer course on the sociopolitical dimensions of climate change.

Emily Twigg, M.S., is an experienced program manager who has worked on a broad range of ocean science topics. From 2016 to 2023, Ms. Twigg served as a program officer for the Ocean Studies Board of the National Academies of Sciences, Engineering, and Medicine. In this capacity, she led or contributed to production of 13 reports and workshop proceedings by coordinating committees of subject matter experts and staff teams, organizing public meetings, building consensus, writing and editing reports, and contributing to public communication materials. She has covered topics ranging from ocean observing, offshore wind development, fisheries, ocean pollution, carbon dioxide removal, and ecosystem restoration. Ms. Twigg also served as the staff officer supporting the U.S. National Committee for the Scientific Committee on Oceanic Research, which is constituted by the National Academies' Ocean Studies Board. She previously held positions at the National Science Foundation supporting multimillion dollar grant programs and the U.S. Environmental Protection Agency supporting an interagency initiative on coastal wetland protection. She has additional experience working in resource management in the National Park Service, and in outdoor environmental education. Ms. Twigg has a master's degree in environmental science and management from the Bren School of Environmental Science and Management at the University of California, Santa Barbara, and a bachelor's degree in biology from the University of California, Berkeley.

Juan Sandoval is a senior program assistant/research assistant with the GRP Division supporting the Gulf Futures Initiative. In his 4 years with the GRP, Mr. Sandoval has provided administrative support on a number of projects, grant cycles, and boards, including the Gulf Health and Resilience Board. Prior to joining the GRP, he worked for the University of Maryland Medical System as a project coordinator. He earned his B.S. in geography from the University of Maryland with a concentration in sustainability and development and a Certificate in Latin American Studies.